# Rock Damage and Fluid Transport, Part II

Edited by
Arno Zang
Ove Stephansson
Georg Dresen

2006

Birkhäuser Verlag
Basel · Boston · Berlin

Reprint from Pure and Applied Geophysics
(PAGEOPH), Volume 163 (2006) No. 10

Editors

Arno Zang
GeoForschungsZentrum Potsdam
Telegrafenberg D425
D-14473 Potsdam
Germany

e-mail: zang@gfz-potsdam.de

Georg Dresen
GeoForschungsZentrum Potsdam
Telegrafenberg D425
D-14473 Potsdam
Germany

e-mail: dre@gfz-potsdam.de

Ove Stephansson
GeoForschungsZentrum Potsdam
Telegrafenberg D425
D-14473 Potsdam
Germany

e-mail: ove@gfz-potsdam.de

Library of Congress Control Number: 2006935064

Bibliographic information published by Die Deutsche Bibliothek:
Die Deutsche Bibliothek lists this publication in the Deutsche Nationalbibliographie;
detailed bibliographic data is available in the internet at <http://dnb.ddb.de>



ISBN 3-7643-7993-6 Birkhäuser Verlag, Basel – Boston – Berlin


Printed on acid-free paper produced from chlorine-free pulp

ISBN 3-7643-7993-6
ISBN 978-3-7643-7993-3

ISBN 978-3-7643-8124-0 (eBook)

9 8 7 6 5 4 3 2 1

Contents

Pure appl. geophys. 163 (2006) 2049–2050
0033–4553/06/102049–2
DOI 10.1007/s00024-006-0114-2



Pure and Applied Geophysics

# Rock Damage and Fluid Transport, Part II

ARNO ZANG, OVE STEPHANSSON, and GEORG DRESEN

## *Introduction*

This volume forms the second part of a topical issue generated from contributions presented at the Euroconference on Rock Physics and Geomechanics that was held at the GeoForschungsZentrum Potsdam, Germany in September 2004. Part I of this volume (Pure and Applied Geophysics, Volume 163, No. 5) mainly contains papers related to problems of rock damage, fracture mechanics, and stresses in rock. Part II contains contributions on transport properties of fractured rocks which were studied with advanced techniques at the laboratory scale and the field scale. *Elsworth and Yasuhara* conduct circulation experiments on fractures in novaculite and limestone, intermittently imaged by X-ray computer tomography. *Xiao et al.* measure permeability, storage capacity, volumetric strain and creep during the deformation of hot-pressed calcite samples with different volume fractions of quartz. *Song and Renner* investigate the influence of heterogeneities on hydraulic properties of Fontainebleau sandstone applying harmonic pore pressure testing. On a larger scale, *Albrecht and Mansurov* estimate porosity and permeability analysing the velocity dispersion of elastic waves near boreholes. *Huenges et al.* monitor fluid pressure variations in a borehole sunk in geothermal reservoir rocks of the Rotliegend formation in NE-Germany.

Three manuscripts from part II focus on rock damage. At the laboratory scale *Ganne and Vervoort* study the difference in crack damage induced by either compressive or tensional stresses in limestones. *Hall et al.* monitor fracture propagation in Neapolitan tuff using acoustic emission and crack imaging techniques to understand collapse phenomena of underground cavities below the city of Naples. *Vinciguerra et al.* report laboratory P- and S-wave velocity measurements on tuff aiming to unravel the seismic velocity structure of the Campi Flegrei Caldera in Italy.

---

GeoForschungsZentrum Potsdam, Telegrafenberg, 14473 Potsdam, Germany.

*Heffer and King* address the question of a metastable or critical crust and the evolution of effective moduli towards the critical point of fracturing. *German* uses a statistical model describing energy scaling of inter-event times (seismic waiting time distributions).

Finally, we would like to express our sincere thanks to the reviewers that contributed their time and efforts to the success of Part II of this volume: *M. Albrecht, M. Bohnhoff, E. Brueckl, M. Di Marzio, D. Elsworth, S. Fomin, J. Fortin, J. Fredrich, A. Gray, V. German, B. Haimson, S. Hainzl, A. Kaselow, P. Knoll, I. Main, H. Milsch, J. Napier, A. V. Nikolaev, J. Renner, F. Roth, K. Sato, A. Schubnel, S. Stanchits, J. ter Heege, B. Thompson, C-F. Tsang, D. Turcotte and R. Zimmermann.*

Pure appl. geophys. 163 (2006) 2051–2070
0033–4553/06/102051–20
DOI 10.1007/s00024-006-0113-3



**Pure and Applied Geophysics**

# Short-Timescale Chemo-Mechanical Effects and their Influence on the Transport Properties of Fractured Rock

Derek Elsworth[1] and Hideaki Yasuhara[1,2]

*Abstract*—Anomalous changes in permeability are reported in fractures circulated by fluids undersaturated with respect to the mineral host. Under net dissolution and net removal of mineral mass, fractures may alternately gape or seal, depending on the prevailing mechanical and chemical conditions. The influence on transport properties is observed to be large, rapid, and irreversible: Permeabilities may change by two orders of magnitude in a month, and the direction of permeability change may switch spontaneously, for no apparent change in environmental forcing. These behaviors are apparent in continuous circulation experiments conducted on fractures in novaculite and limestone, intermittently imaged by X-ray CT. In novaculite, permeability *reduces* by two orders of magnitude as silica is *net removed* from the sample. Surprisingly, these changes can occur at modest temperatures (~80°C) and stresses (~3.5 MPa), where compaction progresses as temperatures are incremented. Isothermal (~20°C) circulation tests in limestone show similar compaction driven by pressure solution. Where circulation remains undersaturated in *Ca*, the change in permeability spontaneously switches from net reduction to net increase as a wormhole forms. The surprising magnitude and rapidity of these changes are investigated in the context of the competition between stress- and chemistry-mediated effects.

**Key words:** Precipitation, dissolution, permeability, fracture, pressure solution.

## 1. *Introduction*

The important interaction between mechanically-driven and chemically-driven processes has been recognized in controlling the evolution of the mechanical and transport constitutive relations for fractured rocks. These interactions are especially important in fractured rocks, where both mechanical stiffness and strength, and permeability, are sensitive to small changes in the infilling or erosion of fracture voids, or changes in contact area. Despite this importance, views of these complex processes have grown in two separate communities; One where mechanical effects dominate, and chemical influences are merely overprinted on these mechanical effects; and one where chemical effects dominate, and mechanical influences are merely overprinted on

[1] Department of Energy and Geo-Environmental Engineering, Penn State University, University Park, PA 16802-5000, U.S.A. E-mail: elsworth@psu.edu
[2] Now at: Department of Civil and Environmental Engineering, Ehime University, Matsuyama, 790-8577, Japan.

these chemical effects. Largely absent have been the roles of mechanically-mediated chemical effects, and their corollary, chemically-mediated mechanical effects.

A limited understanding of mechanisms is available where the full interaction of mechanical and chemical effects is included. The controlling processes and feedbacks apply, in varying degree, to a variety of disciplines, and at a variety of spatial and temporal scales. These include our understanding of diagenetic processes at basin scale, and our design of successful acidizing treatments, geothermal completions, and *in situ* leaching sweeps, at more modest length scales. They include our understanding of short-term transients, the migration of boiling zones, and the development of caprocks in geothermal reservoirs, and relate to biofouling and clogging in shallow aquifers. They apply to the deposition and concentration of ores, as well as our ability to recover these valuable commodities by solution mining. The processes and feedbacks apparent in these complex systems are wide ranging, and have broad significance. Key issues relate to both the strength and the sense of various feedbacks and their influence on mechanical and transport behavior.

Data on the nature of full coupling between mechanical and chemical effects in rock fractures are sparse, although specific results are available for the behavior of natural and artificial fractures in granite at moderate temperatures ($>300°C$) (MOORE *et al.*, 1994), and in volcanic tuff (LIN *et al.*, 1997) and novaculite (POLAK *et al.*, 2003) at lower temperatures (50°–150°C). Related data are also available for composite aggregates of quartz (ELIAS and HAJASH, 1992; YASUHARA *et al.*, 2004), halite (GRATIER, 1993), calcite (ZHANG *et al.*, 1994) and albite (HAJASH *et al.*, 1998), at moderate temperatures (20–120°C) and the same material suites at elevated temperatures and pressures (e.g., ZOBACK and BYERLEE, 1975; SIDDIQI *et al.*, 1997).

Apparent from these experimental data are the competing roles of processes that increase pore or flaw connectivity (NGWENYA *et al.*, 2003), including dilatant shear (ZHANG *et al.*, 1998; ZHANG *et al.*, 1999), microcracking, thermal cracking and focused dissolution (MOK *et al.*, 2002; COLON *et al.*, 2004), and those that destroy or counteract the evolution of pore interconnectivity (AHARONOV *et al.*, 1998; TENTHOREY *et al.*, 1998; MAIN *et al.*, 2000), including shear (ZHU and WONG, 1997) and hydrostatic compaction, fracture healing, dislocation creep, and pressure solution (BOS *et al.*, 2000) prompted by water-film and free-face diffusion (YASUHARA *et al.*, 2004). The sense of permeability change is controlled by the dominant suite of processes; this is in turn largely controlled by the evolving conditions of effective stress, and temperature in the sample, and species concentrations within the effusing fluid.

The more limited studies on fractures (MOORE *et al.*, 1994; LIN *et al.*, 1997; DURHAM *et al.*, 2001; POLAK *et al.*, 2003) amplify these observations, indicating an increased sensitivity to HMC (hydro-mechanical-chemical) processes, even at temperatures as low as 100°C, where the mobile phase is silica, and the test duration is of the order of a month (ELIAS and HAJASH, 1992; LIN *et al.*, 1997), where permeability may be reduced by a factor of 10,000 times (LIN *et al.*, 1997). Recent experimental results involving the circulation of hydrothermal fluids (20°C–150°C) through a natural fracture in novaculite (POLAK *et al.*, 2003; YASUHARA *et al.*, 2004) have also shown reductions in permeability of over two orders-of-magnitude within the period of one month, and of spontaneous switching between permeability reduction and permeability increase as only temperatures and flow rates were changed. These observations are interpreted to result from the spontaneous switching between a regime in which asperity dissolution by pressure solution is dominant, to one where free-face etching and fracture widening dominates. This enigmatic switching in permeability occurs for only subtle changes in temperatures and flow rates, with all other test conditions constant. For normal loading of a fracture, the contributing processes are ill-defined, and anticipated senses and magnitudes of the effects on permeability remain poorly constrained. For shear loading, comparable characterizations that identify key processes and define the expected sense and magnitude of permeability changes and their evolution in time remain ill-defined for porous rocks (BOS *et al.*, 2000; Evans *et al.*, 1997; BERNABE *et al.*, 2003b) and unavailable for fractures.

In this work we describe these embryonic characterizations of the role of chemo-mechanical effects on the evolution of fracture transport properties. These characterizations explain the enigmatic observation that fractures gape with net dissolution (DURHAM *et al.*, 2001; POLAK *et al.*, 2003). Specifically in this work, process-based models are applied to quantify rates of fracture closure observed in prior experimental data where the role of free-face dissolution is shown important (e.g., the experimental results of POLAK *et al.*, 2004).

## 2. *Experimental Observations*

Flow-through experiments are reported for fractured samples confined within an X-ray transparent cell to evaluate the influence of stress-mediated changes in dissolution and precipitation on the evolution in transport properties. Cylindrical samples containing a single axial fracture are confined within a flexible membrane and end-to-end flow tests completed, as illustrated in Figure 1. Three independent measurements are made to constrain the processes promoting the redistribution of mineral mass within the fracture: hydraulic flux, mineral mass flux, and volumetric imaging by X-ray CT. First, permeability is monitored continuously by prescribing flow rate and by measuring pressure drop across the sample. Effective stresses are retained within a narrow range by controlling the upstream back pressure. This measurement records permeability, and through this defines change in average aperture within the system (PIGGOTT and ELSWORTH, 1993). Second, the net efflux of dissolved mineral mass is measured periodically to provide a record of net mass removal, and to correlate this with observed changes in aperture, defined by the flow tests.

Finally, periodic nondestructive imaging by X-ray CT is used to view redistribution of mineral mass within the heated sample, as illustrated in Figure 2. X-ray CT

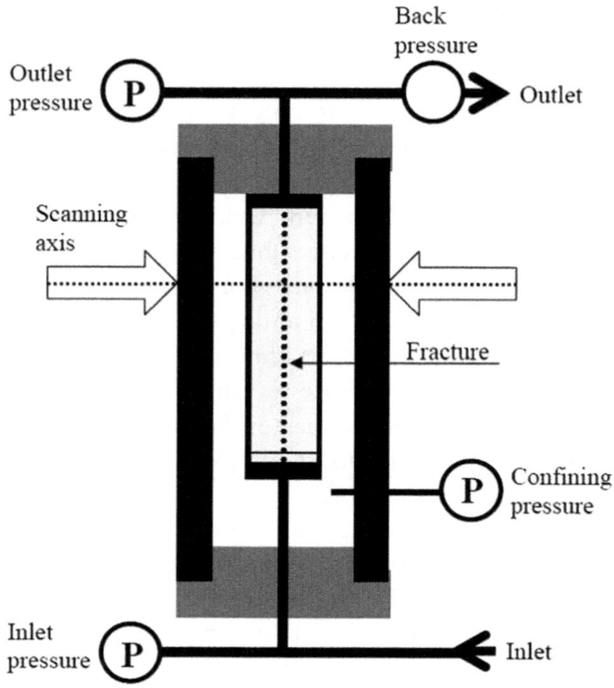


Figure 1
Experimental arrangement showing confining cell and sample with axial fracture. Flow at prescribed flow-rate and measured pressure drop records permeability. Constraint on processes is provided by measurement of dissolved mass efflux, and periodic imaging by X-ray CT (POLAK *et al.*, 2003).

Figure 2
Experimental arrangement showing pressure cell within X-ray CT including plumbing for flow-through tests. Heater tapes provide elevated temperatures to insulated sample.

records changes in density within the sample as a proxy for mineral removal or redistribution, with the scanner used in this study capable of resolving down to ~1/1000th of the diameter of the sample. For sample diameters of 3–5 cm used in the following study, the minimum voxel size scales to the order of 30–50 μm. This resolution is at the limits of utility in defining mass redistribution.

These measurements provide unusual constraint on the evolving processes. Importantly, they allow the source of dissolved components to be determined: We need to discriminate whether the source is from free-face dissolution of the fracture wall, or from stress-mediated dissolution at contacting asperities. This distinction is crucial since these two mechanisms impart opposite effects in the sense of permeability change, under net dissolution: Free-face dissolution increases permeability, and pressure solution reduces permeability. These behaviors are examined under a uniform framework for novaculite and limestone, where the fundamental material properties control response, as identified in Table 1.

### *2.1. Observational Data – Siliceous Rocks*

Constant stress flow-through experiments on a natural fracture in Arkansas Novaculite are reported which observe the bounds of this curious phenomenon—where net dissolution reduces fracture permeability. Pretest scans of the sample, by X-ray CT with a resolution of 37 *μ*m, recover an image of the void space present within the fracture, as illustrated in Figure 3. The "islands" of imaged fracture void are presumed connected via open conduits that are less than 37 μm in aperture, and are eliminated in the thresholding of the image. This is confirmed by the flow tests, initiated under a confining stress of 3.5 MPa at 20°C, where the initial mean hydraulic aperture is of the order of 12 μm, as illustrated in Figure 4. Fluid flow through the sample is at constant rate, $Q$, of 2 cc/min as the temperature is incremented and held, first to 80°, then to 120°, then to 150°C. Apparent during heating is the rapid reduction in aperture, and amplified reduction in fracture permeability that accompanies the incremental stages of heating (POLAK *et al.*, 2003). Measured aperture initially drops precipitously, and then slows to an equilibrium

Table 1

*Contrasting experimental and material characteristics of the component rock types*

| | Novaculite | Limestone |
|---|---|---|
| Matrix porosity | <0.01% | <0.01% |
| Temperature, °C | 2z0–150 | 20 |
| Effective Stress, MPa | 3.5 | 3.5 |
| Permeants | DI | G/W & DI |
| Dissolution rate, $Mol.m^{-2}.s^{-1}$ | $\sim 10^{-9}$ | $\sim 10^{-6}$ |
| Precipitation rate, $Mol.m^{-2}.s^{-1}$ | $\sim 10^{-7}$ | |

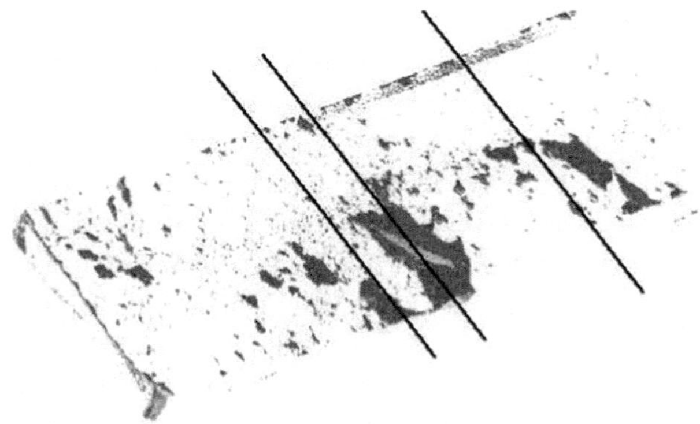

Figure 3
Reconstruction of the fractured sample from X-ray CT imaging [courtesy A.S. Grader].

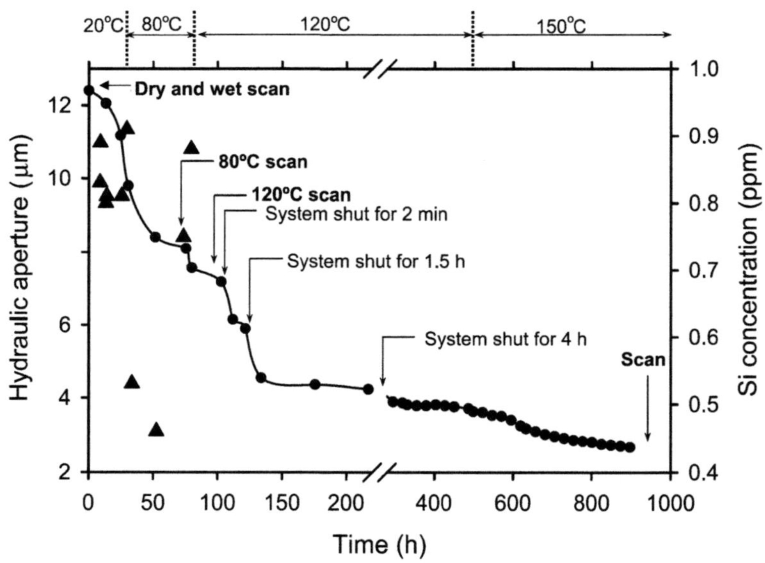


Figure 4
Change in hydraulic aperture with time for a circulation test on a fracture in novaculite. Test is conducted at incremented temperatures but constant stress (POLAK *et al.*, 2003).

magnitude, and occurs with net dissolution of *Si* of the order of 0.77 ppm throughout. These changes in fracture aperture are large, reducing aperture five-fold, and permeability by two orders-of-magnitude, and are surprisingly rapid—each thermally driven cycle completes in a few days, and is complete for the full range of temperatures in only 40 days.

Perhaps even more interesting than the large and rapid change in permeability, is the sense of the permeability change—permeability reduces despite net dissolution from the fracture. The presumed mechanism for this is through pressure solution at the contacting asperities. Locally amplified stresses at asperity contacts elevate the chemical potential of the silica that, in turn, is mobilized into the saturated fracture void, and removed from the system. The reasonableness of this mechanism may be evaluated by comparing the observed change in fracture aperture, $b$, with the anticipated mass removal rate, $dM/dt$, measured in the experiment. For a defined fracture contact area, $A_c$, and density of quartz, $\rho_g$, these are linked through the effluent concentration, $C_p$, as $db/dt = dM/dt \cdot 1/A_c\rho_g = QC_p/A_c\rho_g$ (YASUHARA *et al.*, 2004). For the average effluent concentration of *Si*, of 0.77 ppm and fractional contact areas in the range 5–30%, this results in calculated aperture closure rates of the order of $10^{-10}$ to $10^{-12}$ m/s, with measured closure rates, in the range $10^{-11}$ to $10^{-13}$ m/s. Correspondingly, removal of mineral mass from the contacting asperities is a plausible explanation for the observed behavior, and is consistent with the

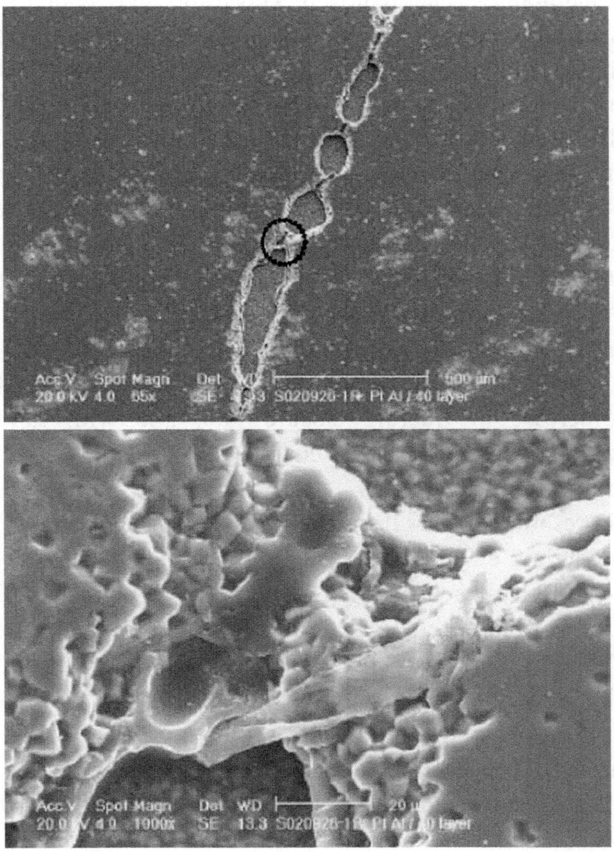


Figure 5
SEM of fracture post test, showing locations of welded fracture [courtesy A.B. Polak].

activation of pressure solution. This behavior is also consistent with observations of the sectioned fracture, post-test, where SEM micrographs show dissolution sites, and the blunting and welding of asperity contacts in a natural fracture that had been separated prior to the test (Fig. 5).

The stress-mediated redistribution of mineral mass by pressure solution may be represented by the three serial processes of: mineral dissolution beneath the contacting asperity, transport along an interfacial water-film to the fracture void, and then redistribution by precipitation within the fracture void, or efflux from the sample. The asperities interpenetrate as mass is removed from contacting asperities, in a similar manner to the mechanisms that drive grain interpenetration and cementation in sandstones (YASUHARA *et al.*, 2003).

### *2.2. Observational Data—Carbonate Rocks*

Flow through tests on fractured carbonates show similar results—permeability net reduces as mineral mass is net removed. This response is again driven by pressure solution. The net (outlet minus inlet) mineral concentrations within fluids circulated during a flow-though test on a fracture in limestone (POLAK *et al.*, 2004) are shown in Figure 6. Groundwater (pH$\sim$8) is circulated for the initial 935 h, followed by the circulation of distilled water (pH$\sim$6) for the final 515 h. The change in permeability recorded through this period is shown in Figure 7. This notes the steady monotonic decline in permeability that occurs with the circulation of groundwater—the fracture net closes despite a small net dissolution. Following the switching of the permeant to distilled water (pH$\sim$6), the reduction in aperture initially accelerates, driven by the amplification of pressure solution as the dissolution rate constant is elevated in the more acidic permeant. This change is manifest as a coherent signal in the elevated output of *Ca*, signalling the more active effect of pressure solution. However, following this increase in fracture closure rate, and with no change in the experimental conditions, permeability change spontaneously switches from net reduction to a rapid increase.

The switch in response results from the observed localization of mass dissolution along a narrow conduit, evident in the fracture morphology captured in CT images taken both pre- and post-test (Fig. 8). These images show the localization of mass removal along a "wormhole" that grows and ultimately dominates flow within the system. Since these experiments provide independent constraint on hydraulic behavior and the redistribution of mass within the sample (through measurements of effluent concentrations and through imaging by X-ray CT), these measurements may be compared. If consistency is sought between the observed hydraulic and mineral mass balance data, then added constraint is applied to the problem, as evident in Figure 7.

Anticipated changes in permeability may be predicted from the mass removal rates, provided an appropriate mechanistic model is prescribed. Assuming that

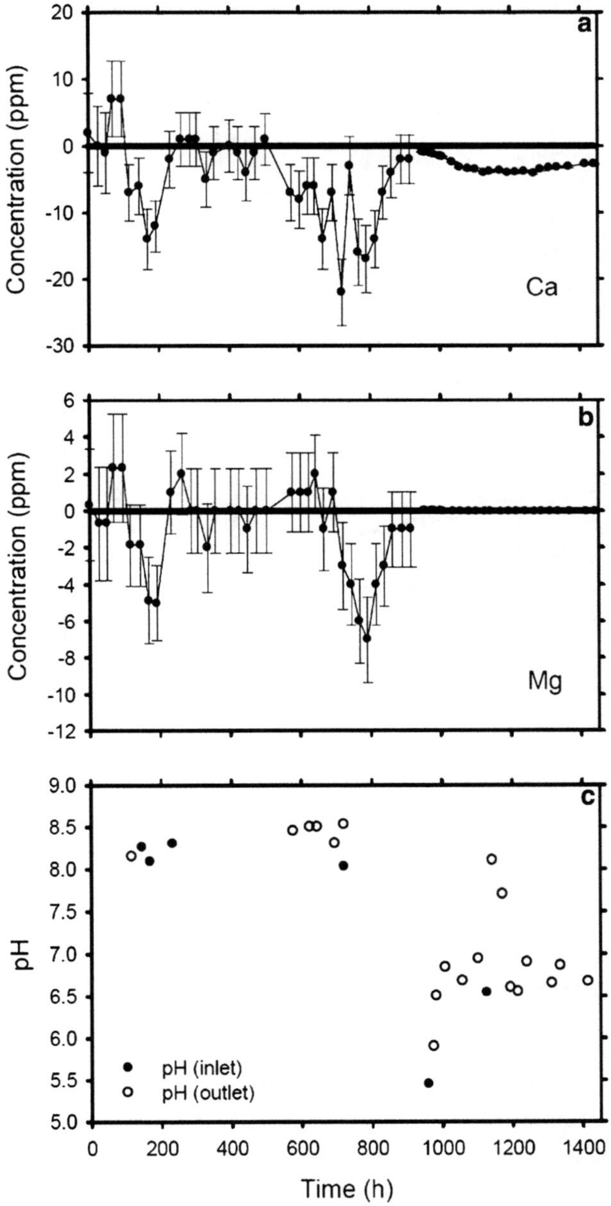


Figure 6
Net efflux in *Ca* and *Mg*, and change in pH with time for the flow-through test on limestone (POLAK *et al.*, 2004).

mass is removed only from the contacting asperities, and that the contact area is typically in the range 5–30%, then permeability may be evaluated directly. Applying this procedure to the data in Figure 7, shows the difficulty in balancing

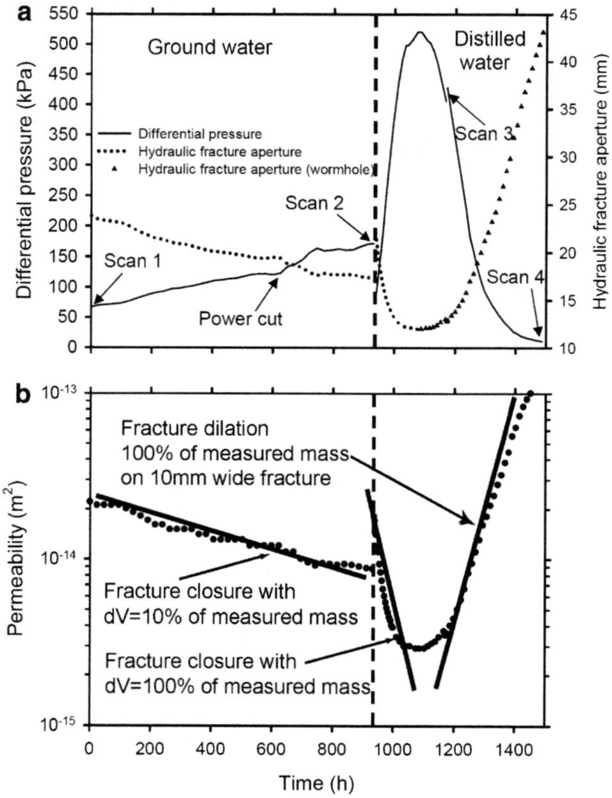


Figure 7
Change in (a) driving differential pressure with time, and for aperture change with time, for flow-through test. (b) Change in permeability with time, evaluated from the flow data (Modified from POLAK *et al.*, 2004).

measured mass removed with that observed in early time—the measured mass removed is a factor of 10 larger than that predicted from the change in permeability, suggesting that large measurement error in the mass concentration data is a major limitation (POLAK *et al.*, 2004). Alternatively, additional dissolved mass from the sample interior, not contributing to the change in aperture could participate in the process, and result in the observed mismatch. Upon switching the influent fluid to distilled water (pH~6), the observed change in permeability closely matches the change predicted from the mass efflux. This matches for an unmeasured but reasonable asperity contact area of 25%. Finally, as the aperture closes, and flow is progressively restricted to the continuous path of the largest initial apertures, one single conduit develops as the principal flow path — the observed mass removal rate is consistent with the evolution of a flow path 10 mm in width. This observation is consistent with the development of the wormhole, apparent in the scanning data of Figure 8. Although only diagnostic, these models

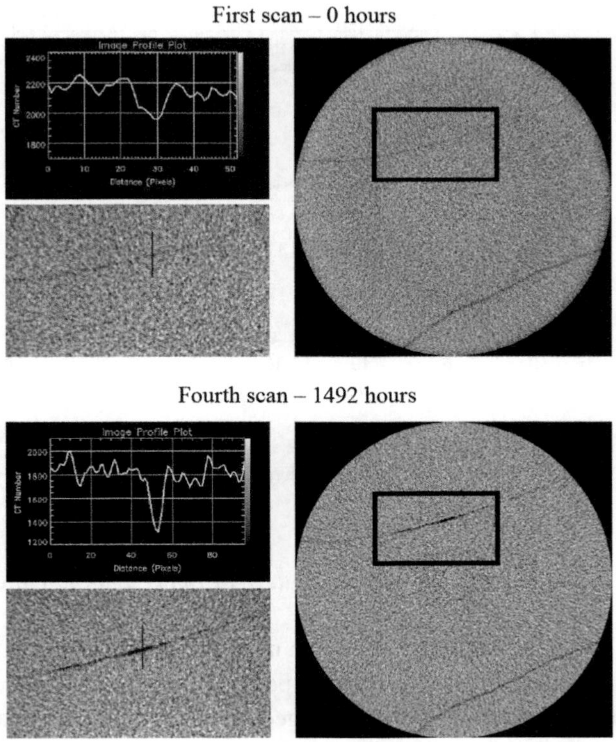


Figure 8
Scan of the fractured limestone sample (above) pre- and (below) post-circulation. Growth of the fracture aperture occurs during the test. Change in CT number across the sample confirms growth of the fracture as a wormhole (POLAK *et al.*, 2004).

are consistent with the findings from prognostic lumped-parameter models that enable this response to be followed *a posteriori* (YASUHARA *et al.*, 2004) and to match proposed mechanisms with observations.

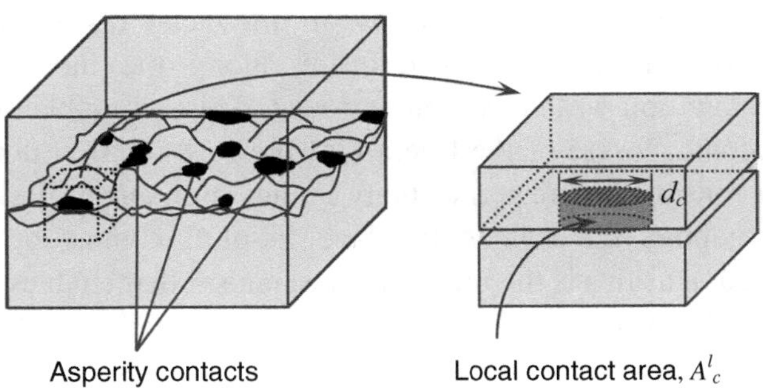


Figure 9
Idealized representation of the asperity contact condition. One representative contact area, $A_c^l$ (right side), represents the area of each asperity in contact (left side), and is considered circular in shape, of diameter, $d_c$.

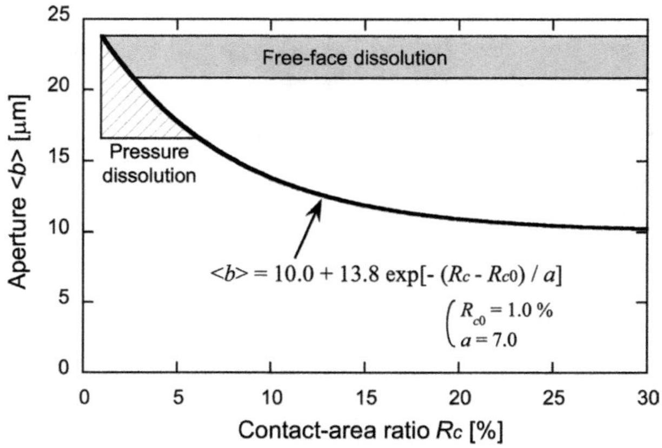


Figure 10
Relationship between aperture and contact area.

## 3. *Micro-Scale Models—Scaling in Space and in Time*

Clear from the experimental data, is the need to scale response in both space and in time. Fracture closure is clearly time dependent, and asymptotes in time to final closure—this in turn is controlled by the geometry of the asperity arrangement, as will be described here.

A mechanistic model is presented to describe the dissolution-dependent evolution of a fracture aperture in rock that accommodates the competing processes of pressure (dis)solution and free-face dissolution. Fracture aperture (or related permeability) may decrease if pressure solution dominates, or may increase if free-face dissolution prevails. Incorporating the role of pressure solution beneath the bridged contact points of a rough fracture has been shown adequate in replicating the observed closure of a natural fracture in novaculite (YASUHARA *et al.*, 2004). This model follows the three serial processes of interfacial dissolution beneath the contacting asperity indentor, mass diffusion along the indentor-iterface, and subsequent precipitation within the fracture void. This is capable of replicating the observed monotonic closure of the fracture as pressure solution dominates (POLAK *et al.*, 2003). In contrast, where permeability change spontaneously switches from net closure to net gaping, the role of free-face dissolution must be accommodated (POLAK *et al.*, 2004), as this is the presumed mechanism that enables the evolution of a dominant etched flow conduit. This development is detailed in the following.

### 3.1. *Rates of Pressure Solution and Free-face Dissolution*

The evolution of fracture aperture is controlled by the competing influences of pressure solution and free-face dissolution. Pressure solution within a fracture incorporates three serial processes; dissolution at asperity contacts, diffusion along the interfacial water film, and precipitation at the pore wall, and may result in net reduction of fracture aperture. This results if the solubility of the solute is above the equilibrium concentration for the solvent. Conversely, if the mass rate of supply to the fluid occupying the fracture void is sufficiently low, or the flow-system sufficiently open, then the solute concentration in the pore fluid will be below the equilibrium concentration, net dissolution will proceed, and the fracture will widen. The competition between pressure solution and precipitation in the fracture void, that together contribute to a net reduction in permeability, and dissolution from the wall of the fracture void, that will net increase permeability, will prescribe the dominant effect; either net sealing or gaping. Importantly, the dominant mechanism may change with stress and the chemical condition of the solvent, or as a result of the evolution of fracture topography, and flow topology.

First, dissolution at the asperity contacts provides a source of mass into the fracture cavity, and is most conveniently defined in terms of a dissolution mass flux, $dM^{PS}_{\mathrm{diss}}/dt$, given as (YASUHARA *et al.*, 2003),

$$\frac{dM^{PS}_{\mathrm{diss}}}{dt} = \frac{3\pi\, V_m^2\, (\sigma_a - \sigma_c)\, k_+ \rho_g\, d_c^2}{4R\,T}, \tag{1}$$

where $V_m$ is molar volume of the solid ($3.69 \times 10^{-5}$ m$^3$ mol$^{-1}$ for calcite), $\sigma_a$ is the disjoining pressure (e.g., HEIDUG, 1995) equal to the amount by which the pressure acting at grain-to-grain contacts exceeds the hydrostatic pore pressure, $k_+$ is the dissolution rate constant of the solid, $\rho_g$ is the grain density (2710 kg m$^{-3}$ for calcite), $d_c$ is the diameter of the asperity contact, $R$ is the gas constant, $T$ is the temperature of the system, and $\sigma_c$ is the critical stress, which defines stress state where the compaction of grain aggregate will effectively halt.

Second, based on Fick's first law and integrating it with respect to a circular contact of radius $r$ in the range $a \leq r \leq d_c/2$, diffusion along the asperity contact area is defined in terms of the diffusive mass flux, $dM_{\mathrm{diff}}/dt$, as (YASUHARA *et al.*, 2003),

$$\frac{dM_{\mathrm{diff}}}{dt} = \frac{2\pi\,\omega\,D_b}{\ln\left(\frac{d_c}{2a}\right)}\left(C_{\mathrm{int}} - C_{\mathrm{pore}}\right), \tag{2}$$

where $\omega$ is the thickness of the water film trapped at the interface, $D_b$ is the diffusion coefficient, and $(C_{\mathrm{int}})_{x=a}$ and $(C_{\mathrm{pore}})_{x=d_c/2}$ are mineral concentrations in the interface fluid and pore space, respectively.

Finally, free-face dissolution/precipitation are quantified as mass fluxes, $dM^{FF}_{\mathrm{diss}}/dt$ and $dM_{\mathrm{prec}}/dt$, defined by the dissolution/precipitation rate constants and difference between the fluid mass concentration in the pore space and the equilibrium concentration, defined as (modified from PALMER, 1991; DREYBRODT and BUHMANN, 1991),

$$\frac{dM_{\text{diss}}^{FF}}{dt} = k_+ A_{\text{pore}} \rho_g V_m \left(1 - \frac{C_{\text{pore}}}{C_{\text{eq}}}\right)^n, \tag{3}$$

$$\frac{dM_{\text{prec}}}{dt} = k_- A_{\text{pore}} \rho_g V_m \left(\frac{C_{\text{pore}}}{C_{\text{eq}}} - 1\right)^n, \tag{4}$$

where $A_{\text{pore}}$ is the area of the fracture void, $k_-$ is the precipitation rate constant of the dissolved mineral, and $C_{\text{eq}}$ is the equilibrium solubility of the dissolved mineral. $n$ denotes the reaction order, which is experimentally constrained (DREYBRODT, 1998): The reaction kinetics is roughly first order for a pore fluid undersaturated with calcium, and the reaction order increases as the water approaches saturation. In this preliminary model, the order is set to unity as a fist approximation. Note that the dissolution/precipitation mass fluxes will be zero as the mass concentration in the pore fluid is either greater or smaller than the equilibrium solubility, respectively.

### 3.2. Geometric Effects

A single fracture is idealized as two rough surfaces held apart by bridging asperities, as illustrated in Fehler! 'Verweisquelle konnte nicht gefunden warden.' (left side) The average contact-area ratio, $R_c$, may be determined by defining a representative contact area, $A_c^l$, surrounded by an appropriate tributary area, $A_t^l$, '(Fehler! Verweisquelle konnte nicht gefunden warden'.; right side), and is assumed equivalent to the ratio of the summed local contact areas, $A_c^t$, to the total fracture area, $A_t^t$, given as (YASUHARA *et al.*, 2004),

$$R_c = \frac{A_c^l}{A_t^l} = \frac{A_c^t}{A_t^t}. \tag{5}$$

Within this tributary area, the contact diameter, $d_c$, of the local contact area, $A_c^l$, is defined as,

$$d_c = \sqrt{\frac{4A_c^l}{\pi}}. \tag{6}$$

For uniaxial compaction, the normal forces acting on the tributary area and the contacting asperity balance, yielding the stress applied at the contact area, $\sigma_a$, as,

$$\begin{aligned} \sigma_{\text{eff}} \cdot A_t^l &= \sigma_a \cdot A_c^l \\ \Rightarrow \sigma_a &= \frac{\sigma_{\text{eff}}}{R_c} \end{aligned}, \tag{7}$$

where $\sigma_{\text{eff}}$ is the average macroscopic effective stress.

Interactive processes of pressure solution and free-face dissolution irreversibly alter the geometry of the fracture surfaces, and the relation between fracture aperture and contact area may be defined to follow this modification of the fracture aperture and contact-area ratio within the tributary domain. A simple, but physically viable, relation between them is defined as (YASUHARA *et al.*, 2004),

$$\langle b \rangle = a_1 + a_2 \cdot \exp\left(-(R_c - R_{c0})/a_3\right), \tag{8}$$

where $\langle b \rangle$ is the mean aperture, $R_c$ is the contact-area ratio, and $a_i$ ($i = 1, 2, 3$) is a constant. This curve is adopted as a straightforward and representative relation between fracture contact area and aperture, to define the phenomenology of fracture sealing/gaping by pressure solution/free-face dissolution.

### 3.3. Computational Procedure

The individual processes of dissolution at asperity contacts, diffusion along interfacial water film, and free-face dissolution/precipitation are combined to define the progress of aperture reduction of the fracture with time. In the initial condition, a small representative contact area is set with the initial aperture of the fracture. An effective stress is applied, as amplified by the tributary geometry, and during time step $\Delta t$, appropriate magnitudes of mass dissolution at the representative contact area, diffusion, and free-face dissolution/precipitation are simultaneously evaluated from Eqs. (1), (2), (3), and (4), respectively. Physically, the dissolved mass evaluated from Eq. (1) is supplied to the interface, and domain shortening (i.e., aperture reduction) proceeds as this mass passes along the interface by diffusion, as defined by Eq. (2). From the known magnitude of the diffusing mass, the updated contact area and aperture are calculated using the relation of Eq. (14) (the integration of Eq. (8) represents the volume that is removed, and its volume is matched by the diffused volume). A portion of the mass that diffuses to the pore fluid may deposit to the free surface of the fracture (Eq. (4)), resulting in an additive reduction in fracture void volume. Alternately, net dissolution from the fracture wall (Eq. (3)) and resulting enlargement of the void cavity will compete with the closure occasioned by the shortening of the bridging asperity. The dominant process will prescribe whether the fracture gapes or seals '(Fehler! Verweisquelle konnte nicht gefunden warden.)'. This deposition or dissolution on the free surface is controlled by the relative concentration differential between the pore fluid solution and the equilibrium concentration of that fluid (Eqs. (3) and (4)). Concurrently, mineral concentrations in the immobile fluid layer beneath the asperity contact, and the mobile fluid in the fracture void fluid are updated, as (YASUHARA *et al.*, 2004),

$$C_{\text{int}}\big|_{t+\Delta t} = \frac{\left(D_1 + V_p/2\Delta t\right) \cdot \left(dM_{\text{diss}}/dt + V_p/4\Delta t \cdot C_{\text{int}}\big|_t\right) + D_1 V_p/2\Delta t \cdot C_{\text{pore}}\big|_t}{\left(D_1 + V_p/4\Delta t\right) \cdot \left(D_1 + V_p/2\Delta t\right) - D_1^2}, \tag{9}$$

where,

$$D_1 = \frac{2\pi\varpi D_b}{\ln\left(d_c/2a\right)}, \tag{10}$$

$$C_{\text{pore}} = \frac{1}{Q}\left(\frac{dM_{\text{diff}}}{dt} + \frac{dM_{\text{diss}}^{FF}}{dt}\right) \quad \left(\text{if} \quad C_{\text{pore}} < C_{\text{eq}}\right), \tag{11}$$

$$C_{\text{pore}} = \frac{1}{Q}\left(\frac{dM_{\text{diff}}}{dt} - \frac{dM_{\text{prec}}}{dt}\right) \quad \left(\text{if} \quad C_{\text{pore}} > C_{\text{eq}}\right), \tag{12}$$

and $Q$ denotes the flow rate.

These relations are used iteratively to follow the evolution of dissolved concentrations within the fracture void, and resulting closure history of the fracture.

## 4. Match with Observed Behavior

The evolution of fracture aperture with the progress of the experiment is represented by the prior model. We follow the measured change in aperture observed in the experiment, from an initial aperture of 24 μm; first closing, and then spontaneously gaping (Polak *et al.*, 2004). Importantly, we provide no *ad hoc* mechanism to trigger the spontaneous change to gaping; rather this evolves as a natural consequence of the competition between agents that promote either asperity removal, or alternately free-face etching, explicitly represented in the model.

The predicted rates of evolutions in aperture and differential concentration in Ca (outlet minus inlet) are shown in Figure 11, together with the experimental measurements. This response is derived for the physical parameters identified in Table 2, but modified in magnitude by the scaling parameters previously described.

*Stage I Behavior (pH~8):* With the fractional contact area initiated at 1%, the evolving modeled aperture closely matches that evaluated from the flow data. The fracture evolution is insensitive to this choice, and reaches an intermediate contact area of 6% at the close of stage I (932 hr), corresponding to an aperture of 18 $\mu$m (Fig. 11(a)). Measured and predicted *Ca* concentrations are compared in Figure 11(b). Evaluated concentration changes (outlet minus inlet) are of the order ~0.01 ppm, and are absent the broad swings recorded in the experimental measurements (Polak *et al.*, 2004), attributed to large relative measurement errors, and net dissolution from the matrix, unaccounted in the model. Predicted *Ca* concentrations in the pore fluid are slightly larger than the equilibrium solubility, which results in nominal precipitation to the pore wall; the diminution in fracture aperture is less than ~10 nm, and is indiscernible in Fig. 11. Correspondingly, aperture evolution is controlled only by stress-mediated dissolution.

*Stage II and III Behavior (pH ~ 6):* As distilled water is introduced, the fracture closure initially accelerates, and the contact area increases to ~15% at the minimum aperture of 12 μm (1082 hr). The closure accelerates due to the increase in stress-mediated dissolution (Eq. (13)) prompted by the lowering of pH in the influent fluid,

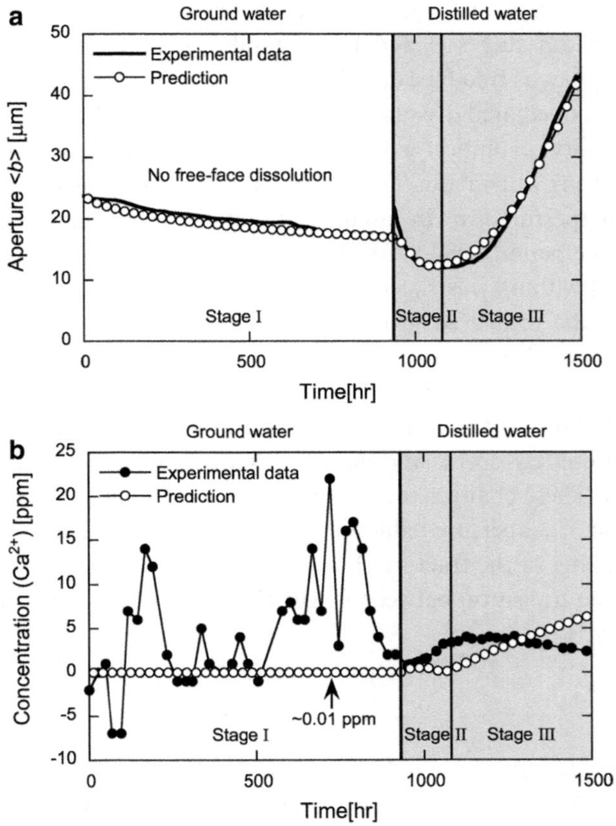


Figure 11
Comparisons between experimental data of POLAK *et al.* (2004) and predictions of the current model, (a) aperture change with time and (b) change in calcium concentrations (outlet minus inlet) with time. Open circles show the predictions using modified values; with $k_+ = 0.01 \times (1.15 \times 10^{-6})$ mol m$^{-2}$ s$^{-1}$ in stage I; with $k_+ = 2.0 \times (1.15 \times 10^{-6})$ mol m$^{-2}$ s$^{-1}$ in stages II and III, and $A_{\text{pore}}$ is decreased by a factor of 1/200 in stages I and II and the factors of $A_{\text{pore}}$ are linearly increased from 1/200 to 60/200 in stage III.

Table 2

*Parameters used to represent the experimental results*

| Temperature $T$ [°C] | Effective stress $\sigma_{\text{eff}}$ [MPa] | Critical stress $\sigma_c$ [MPa] | Diffusion path Width $\omega$ [nm] | Diffusion coefficient $D_b$ [m$^2$s$^{-1}$] | Dissolution rate constant $k_+$ [mol m$^{-2}$s$^{-1}$] | Precipitation rate constant $k_-$ [mol m$^{-2}$s$^{-1}$] | Solubility of calcium $C_{\text{eq}}$ [ppm] |
|---|---|---|---|---|---|---|---|
| 20 | 3.0 | 20.0 | 4.0 | $6.93 \times 10^{-10}$ | $1.15 \times 10^{-6}$ | 1.78 | 37.4 |

as reflected in the elevated dissolution constant, $k_+$. During stage II, stress-mediated dissolution dominates, but this dominance switches to free-face dissolution as the fracture topography and flow topology chang in response to dissolution. This switch is marked as the onset of stage III, but importantly, experimental variables are invariant throughout stages II and III. If model parameters are not changed, the fracture slowly gapes as free-face dissolution builds in the constricted void, relative to the effect of stress-mediated dissolution. This spontaneous switching in aperture, and related permeability, evolution is illustrated in Figure 11(a), although the modeled rate of aperture growth in this final stage is muted over observed behavior. The observed rate of aperture growth can only be matched if the surface area is arbitrarily increased over the period 1082 to 1487 hours, to represent the observed localization of mass removal within a restricted conduit. In this instance, the modeled response closely matches the measured growth in effective aperture from ~12 to ~43 $\mu$m, although the assumed geometry of the fracture remains parallel-sided rather than tubular in section (POLAK *et al.*, 2004).

Throughout stage II, the contact-area ratio monotonically increases from ~6 to ~15%, proportionately decreasing the contact stresses, $\sigma_a$, and increasing the mass diffusion length. These changes provide a reinforcing feedback in slowing the rate of pressure solution (i.e., aperture reduction), allowing free-face dissolution to dominate, and promote gaping of the fracture. For this sample, this transition occurs at ~1028 hr, and marks the transition between stages II and III of the experiment.

## *5. Conclusions*

The previous has documented the surprising influence that complex paths of temperature and chemical potential may exert on the transport properties of fractured rocks. Pressure solution is shown to be potentially an important factor, which, under certain circumstances may strongly influence response. Pressure solution, representing a mechanism of stress-mediated (enhanced) dissolution, can be mobilized at relatively modest temperatures (~20°–80°C) and stresses (~3.5 MPa), and can act rapidly — on the order of a few to tens of days. These effects are shown to be surprisingly virulent — capable of changing permeabilities by many orders of magnitude within these short time frames. Surprisingly, these effects may result in the net reduction of fracture aperture, and consequent reduction in permeability, even when mineral mass is net dissolved and removed from the sample, throughout. Correspondingly, these processes may exert a significant influence on the circulation of fluids in natural and engineered environments, especially those involving significant temporal or spatial gradients in stress, chemical potential, or temperature, and warrant close scrutiny.

## *Acknowledgements*

This work has been supported by grants DOE-BES-DE-FG02-00ER15111, DOE-DE-PS26-01NT41048, DOE-DE-FG36-04GO14289, and ARC DP0209425.

This support is gratefully acknowledged. This work would not have been possible without the contributions of Amir Polak, Avrami Grader, Phil Halleck, and Jishan Liu. Mike Howard and Charles Stone (Arkansas Geological Commission) are thanked for providing a bulk sample of Arkansas novaculite. The comments of Sergei Fomin and one anonymous reviewer are appreciated.

## References

AHARONOV, E., TENTHOREY, E., and SCHOLZ, C.H. (1998), *Precipitation sealing and diagenesis – 2. Theoretical analysis,* J. Geophys. Res.-Solid Earth *103*, 23,969–23,981.

BERNABE, Y., MOK, U., and EVANS, B. (2003), *Permeability-porosity relationships in rocks subjected to various evolution processes*, Pure Appl. Geophys. *160*, 937–960.

BOS, B., PEACH, C.J., and SPIERS, C.J. (2000), *Slip behavior of simulated gouge-bearing faults under conditions favoring pressure solution*, J. Geophys Res.-Solid Earth *105*, 16,699–16,717.

COLON, C.F.J., OELKERS, E.H., and SCHOTT, J. (2004), *Experimental investigation of the effect of dissolution on sandstone permeability, porosity, and reactive surface area*, Geochim. et Cosmochim. Acta *68*, 805–817.

DREYBRODT, W. and BUHMANN, D. (1991), *A mass transfer model for dissolution and precipitation of calcite from solutions in turbulent motion*, Chem. Geol. *90*, 107–122.

DREYBRODT, W., *Principles of karst evolution from initiation to maturity and their relation to physics and chemistry*. In *Global Karst Correlation* (D.X. Yuan, and Z. H. Liu, eds.) (Science Press, Beijing, (1998)) pp. 33–49.

DURHAM, W.B., BOURCIER, W.L., and BURTON, E.A. (2001), *Direct observation of reactive flow in a single fracture*, Water Resour. Res. *37*, 1, 1–12.

ELIAS, B. P. and HAJASH, A. (1992) *Changes in Quartz Solubility and Porosity Due to Effective Stress - an Experimental Investigation of Pressure Solution.* Geology *20* (5) 451–454.

EVANS, J.P., FORSTER, C.B., and GODDARD, J.V. (1997), *Permeability of fault-related rocks, and implications for hydraulic structure of fault zones*, J. Struct. Geology *19*, 1393–1404.

GRATIER, J.-P. (1993), *Experimental pressure solution of halite by an indenter technique*, Geophys. Res. Lett. *20*, 1647–1650.

HAJASH, A., Carpenter, T.D., and Dewars, T.A. (1998), *Dissolution and time-dependent compaction of albite sand: Experiments at 100*°C and 160°C in pHpH-buffered organic acids and distilled water, Tectonophysics *295* (1-2), 93–115.

HEIDUG, W.K. (1995), I*ntergranular solid-fluid phase transformations under stress: The effect of surface forces*, J. Geophys. Res. *100*, 5931–5940.

LIN, W., ROBERTS, J., GLASSLEY, W., and RUDDLE, D. (1997), *Fracture and matrix permeability at elevated temperatures*, Workshop on Significant Issues and Available Data. Near-Field/Altered-Zone Coupled Effects Expert Elicitation Project, San Francisco, November.

MAIN, I.G., KWON, O., NGWENYA, B.T., and ELPHICK, S.C. (2000), *Fault sealing during deformation-band growth in porous sandstone*, Geology *28*, 1131–1134.

MOK, U., BERNABE, Y., and EVANS, B. (2002), *Permeability, porosity and pore geometry of chemically altered porous silica glass*, J. Geophys. Res.-Solid Earth *107*.

MOORE, D.E., LOCKNER, D.A. et al. (1994), *Reduction of Permeability in Granite at Elevated-Temperatures.* Science *265* (5178) 1558–1561.

NGWENYA, B.T., KWON, O., ELPHICK, S.C., and MAIN, I.G. (2003), *Permeability evolution during progressive development of deformation bands in porous sandstones*, J. Geophys. Res.-Solid Earth *108*.

PALMER, A.N. (1991), *Origin and morphology of limestone caves*, Geol. Soc. Am. Bull. *103*, 1–21.

PIGGOTT, A.R. and ELSWORTH, D. (1993), *Laboratory assessment of the equivalent apertures of a rock fracture,* Geophys. Res. Lett. *20* (13), 1387–1390.

POLAK, A., ELSWORTH, D., YASUHARA, H., GRADER, A.S., and HALLECK, P.M. (2003), *Permeability reduction of a natural fracture under net dissolution by hydrothermal fluids*, Geophys. Res. Lett. *30* (20), 2020, doi:10.1029/2003GL017575.

POLAK, A., ELSWORTH, D., LIU, J., and GRADER, A. (2004), *Spontaneous switching of permeability changes in a limestone fracture under net dissolution*, Water. Resour. Res. *40*, W03502, doi:10.1029/2003WR002717.

SIDDIQI, G., EVANS, B., DRESEN, G., and FREUND, D. (1997), *Effect of semibrittle deformation on transport properties of calcite rocks*, J. Geophys. Res. *102* (B7), 14,765–14,778.

TENTHOREY, E., SCHOLZ, C.H., AHARONOV, E., and LEGER, A. (1998), *Precipitation sealing and diagenesis – 1. Experimental results*, J. Geophys. Res.-Solid Earth *103*, 23,951–23,967.

YASUHARA, H., ELSWORTH, D., and POLAK, A. (2003), *A mechanistic model for compaction of granular aggregates moderated by pressure solution*, J. Geophys. Res. *108* (B11), 2530, doi:10.1029/2003JB002536.

YASUHARA, H., ELSWORTH, D., and POLAK, A. (2004), *The evolution of permeability in a natural fracture: Significant role of pressure solution*, J. Geophys. Res. *109*, B03204, doi:10.1029/2003JB002663.

ZHANG, S., COX, S., and PATERSON, M. (1994), *The influence of room temperature deformation on on porosity and permeability in calcite aggregates*, J. Geophys. Res. *99*, 15,761–15,775.

ZHANG, S.Q., TULLIS, T.E., and SCRUGGS, V.J. (1999), *Permeability anisotropy and pressure dependency of permeability in experimentally sheared gouge materials*, J. Struct. Geol. *21*, 7, 795–806.

ZHU, W.L. and WONG, T.F. (1997), *The transition from brittle faulting to cataclastic flow: permeability evolution*, J. Geophys. Res. *102*, B2, 3027–3041.

ZOBACK, M. and BYERLEE, J. (1975), *The effect of microcrack dilatancy on the permeability of Westerly Granite* J. Geophys. Res. *80*, 752–755.

(Received February 24, 2005, revised October 28, 2005, accepted November 10, 2005)

Pure appl. geophys. 163 (2006) 2071–2102
0033–4553/06/102071–32
DOI 10.1007/s00024-006-0115-1



Pure and Applied Geophysics

# Permeability Evolution During Non-linear Viscous Creep of Calcite Rocks

Xiaohui Xiao,[1] Brian Evans,[1] and Yves Bernabé[2]

*Abstract*—Permeability, storage capacity and volumetric strain were measured *in situ* during deformation of hot-pressed calcite aggregates containing 10, 20, and 30 wt% quartz. Both isostatic and conventional triaxial loading conditions were used. The tests were performed at confining pressure of 300 MPa, pore pressures between 50 to 290 MPa, temperatures from 673 to 873 K and strain rates of $3 \times 10^{-5}$ s$^{-1}$. Argon gas was used as the pore fluid. The initial porosities of the starting samples varied from 5% to 9%, with higher porosity correlated to higher quartz content. Microstructural observations after the experiment indicate two kinds of pores are present: 1) Angular, crack-like pores along boundaries between quartz grains or between quartz and calcite grains and 2) equant and tubular voids within the calcite matrix. Under isostatic loading conditions, the compaction rate covaries with porosity and increases with increasing effective pressure. Most of the permeability reduction induced during compaction is irreversible and probably owes to plastic processes. As has been found in previous studies on hot-pressed calcite aggregates, permeability, $k$, is nonlinearly related to porosity, $\phi$. Over small changes in porosity, the two parameters are approximately related as $k \propto \phi^n$. The exponent $n$ strongly increases as porosity decreases to a finite value (from about 4 to 6% depending on quartz content), suggesting a porosity percolation threshold. When subjected to triaxial deformation, the calcite-quartz aggregates exhibit shear-enhanced compaction, but permeability does not decrease as rapidly as it does under isostatic conditions. During triaxial compaction the exponent $n$ only varies between 2 and 3. Non-isostatic deformation seems to reduce the percolation threshold, and, in fact, enhances the permeability relative to that at the same porosity during isostatic compaction. Our data provide constraints on the governing parameters of the compaction theory which describes fluid flow through a viscous matrix, and may have important implications for expulsion of sedimentary fluids, for fluid flow during deformation and metamorphism, and melt extraction from partially molten rocks.

**Key words:** Permeability, porosity, viscous creep, compaction.

## 1. *Introduction*

Fluid flow is an important aspect of many tectonic, metamorphic and diagenetic processes. However, to predict the rate of fluid flow in natural conditions is

[1] Department of Earth, Atmospheric, and Planetary Sciences, Massachusetts Institute of Technology, Cambridge, MA 02139, USA. E-mail: xhxiao@mit.edu; brievans@mit.edu
[2] Ecole et Observatoire des Sciences de la Terre de Strasbourg, Université Louis Pasteur, 5, rue René Descartes, 67084 Strasbourg Cedex, France. E-mail: ybernabe@eost.u-strasbg.fr

challenging, because the permeability of rocks may vary both spatially and temporally by over 11 orders of magnitude, owing to diverse mechanical and chemical interactions, and to the wide range of pressure, temperature and stress state conditions throughout the crust (BRACE, 1980). The brittle upper crust seems to have sufficient strength to maintain hydrostatic fluid pressure gradients, provided that an interconnected pore space exists. Below the brittle-ductile transition, permeability might be exceedingly small, due to compaction facilitated by intracrystalline plasticity or viscous creep. The ductile lower crust may consist of isolated domains of relatively high permeability, where the fluid pressures are at or near lithostatic values (CONNOLLY and PODLADCHIKOV, 2004; ETHERIDGE *et al.*, 1984; NUR and WALDER, 1992; WALDER and NUR, 1984). Fluid escape from metamorphic rocks likely involves episodic hydrofracturing (FISHER and BRANTLEY, 1992; NUR and WALDER, 1992) or porosity-wave propagation driven by the difference between the fluid and rock pressure gradients. The latter has been the subject of numerous theoretical and numerical models dealing with the fundamental problem of melt segregation, sediment compaction and metamorphic fluid flow (CONNOLLY and PODLADCHIKOV, 2000; CONNOLLY, 1997; MCKENZIE, 1984, 1987; SCOTT, 1988; SCOTT and STEVENSON, 1984; SPIEGELMAN, 1993; STEVENSON, 1989).

Some important experimental studies have investigated the evolution of rock permeability during brittle faulting and dilatant cataclastic flow of sandstones (HEILAND and RAAB, 2001; HOLCOMB and OLSSON, 2003; MAIN *et al.*, 2000; NGWENYA *et al.*, 2003; OJALA *et al.*, 2003; VAJDOVA *et al.*, 2004; ZHU and WONG, 1997), halite (PEACH and SPIERS, 1996; POPP *et al.*, 2001) and calcite aggregates (FISCHER and PATERSON, 1992; ZHANG *et al.*, 1994a), and fault gouges (FAULKNER and RUTTER, 2000, 2003; UEHARA and SHIMAMOTO, 2004). The permeability evolution is strongly dependent on the deformation mechanism and failure mode (e.g., ZHU *et al.*, 1997). Dilatancy and brittle faulting generally increase permeability while shear-enhanced compaction caused by grain crushing and pore collapse leads to a decrease. Development of shear and compaction bands can also cause permeability reduction. Fewer measurements of permeability change have been made at high temperature and pressure (but, see FISCHER and PATERSON, 1992; ZHANG *et al.*, 1994b), especially during deformation. High-temperature, isostatic and conventional triaxial tests of synthetic calcite marbles containing 10, 20, or 30 wt% quartz with up to 9% residual porosity were reported earlier (XIAO and EVANS, 2003). Since argon gas was used as pore fluid, pressure-solution effects were not important in these experiments, and straining was most probably dominated by dislocation motion, twinning, and/or diffusion creep. In this study, we give an account of the effect of viscous creep on the permeability of these porous rocks during these tests. Moreover, the addition of rigid quartz particles induced higher scales of heterogeneity in the calcite-quartz aggregates (BERNABÉ *et al.*, 2004). Another objective of this study was to assess the importance of permeability heterogeneity and its interplay with deformation at elevated temperature and pressures.

## 2. Experiments

### 2.1. Starting Material

The calcite-quartz aggregates were fabricated from mixtures of powders of reagent-grade calcite (>99 wt% $CaCO_3$ and <0.2 wt% Na and <0.1 wt% Sr, Riedel de Haen, Germany) and crushed quartz (Ottawa sand, U.S. Silica) in a commercial hot isostatic press (HIP) at 700°C and 200 MPa confining pressure for 3 hours at Bodycote IMT, Andover, MA. Cylindrical samples with 10 mm diameter and 20 mm length were cored and ground from three large HIPed blocks, containing 10, 20, and 30 wt% quartz. In these materials trace amounts of wollastonite (i.e., <0.1 vol%) are also present (RYBACKI *et al.*, 2003) but were neglected in the rest of this study.

The bulk density was determined by precisely machining the aggregate cores to right circular cylinders and then measuring their dimensions and weight. The total porosity of the starting samples could then be calculated using the bulk density of the aggregate and the single-crystal densities of calcite and quartz, assuming the quartz content had negligible fluctuation. Interconnected porosity was measured using Archimedes method, i.e., by comparing the sample dry weight to the weight measured while totally immersing the saturated sample in water. Starting total porosities vary from 5 to 9%, higher porosity being correlated with higher quartz content (Table 1).

Observation of polished sections under the scanning electron microscope (SEM) revealed that the quartz particles generally form clusters of a few grains, randomly dispersed throughout the hot-pressed calcite matrix (Fig. 1). Because the HIP kinetics of quartz are considerably slower than that of calcite, the quartz clusters and their immediate surroundings remained highly porous during HIP, while the bulk of the calcite matrix tightly densified. The average grain size of the aggregates was determined from SEM micrographs using line intercept method, without applying any geometric factor. The mean grain size of both the quartz and calcite grains determined by that method is about the same, on the order of 10 μm, although a number of enlarged calcite grains can be observed, suggesting that grain growth occurred in the calcite matrix. Two kinds of pores can be observed in Figure 1: Relatively large (3 to 6 μm), angular, elongated pores within and around the quartz

Table 1

*Porosity and permeability of the starting material*

| Quartz content (wt%) | Total porosity (%) | Connected porosity (%) | Permeability* ($10^{-18}$ $m^2$) |
|---|---|---|---|
| 10 | 5.4 ± 0.5 | 4.2 ± 0.4 | 0.2 ± 0.05 |
| 20 | 7.6 ± 0.6 | 6.9 ± 0.5 | 2.8 ± 0.3 |
| 30 | 9.2 ± 0.2 | 8.6 ± 0.5 | 7.8 ± 1.5 |

* At temperature of 873 K, and effective pressure of 10 MPa.

clusters and submicron, tubular voids (located at three-grain edges, see BERNABÉ *et al.*, 1982) within the calcite matrix.

BERNABÉ *et al.* (2004) investigated the permeability of these materials at room temperature and in isostatic conditions (with the effective pressure up to about 100 MPa) using water as a pore fluid. They demonstrated that these rocks can be described as a binary mixture of a high- and a low-permeability phase, i.e., the highly porous regions associated with the quartz clusters and the highly densified calcite matrix, respectively. One important result was that the high-permeability regions form a connected path in the 20 and 30 wt% quartz varieties but not in the 10 wt% one (also compare Figs. 1b and 1c to 1a). It was also found that the pressure dependence of permeability was very small in the 10 wt% rock and significantly larger in the other two materials, confirming that the calcite matrix only contains pressure-resistant, tubular pores, whereas the quartz clusters are characterized by pressure-sensitive, crack-like pores.

### 2.2. Procedures

The tests were carried out in a gas-medium deformation apparatus (PATERSON Instruments) with internal furnace, internal load cell and a servo-controlled pore-fluid volumometer. Specimens, split alumina spacers and alumina pistons were inserted in iron jackets, and connected to an upstream and downstream reservoir. An inner copper shell was placed between the specimen and the iron jacket in order to provide more intimate contact around the cylindrical surface of the specimen and reduce the apparent pore volume associated with surface irregularities. The pore fluid pressures in both upstream and downstream reservoirs were monitored by separate pressure transducers to a resolution of 0.1 MPa. A servo-controlled motor drove the volumometer piston and maintained the pore pressure constant. The pore volume changes in the sample can then be continuously inferred from the piston displacement $\Delta x_{vol}$ recorded by a transducer (LVDT) with a resolution of 0.3 μm, attached to the end of the piston. The resolution of $\Delta x_{vol}$ (0.3 μm) is equivalent to a resolution of about $2 \times 10^{-5}$ in volumetric strain.

During an infinitesimal time interval $dt$, the decrease in connected pore space produced by compaction is given to first order by $dV_c = dV_{cc} - dV_{cu} + dV_{uc}$ while the

▶

Figure 1
Backscattered SEM image of hot-pressed two-phase aggregates with 10 wt% (a), 20 wt% (b) and 30 wt% (c) quartz particles (dark gray) after fabrication, but before additional compaction or deformation. Two kinds of pores can be seen: Cracks along grain and interphase boundaries and more equant voids within the calcite matrix, usually at the junctions of several grains, but also, rarely, included within a single grain. Because of the large contrast in hardness between the two minerals, it is possible that some of the porosity seen at the interphase boundaries results from polishing damage, although care was taken to minimize this effect. Scale bar corresponds to 100 μm.

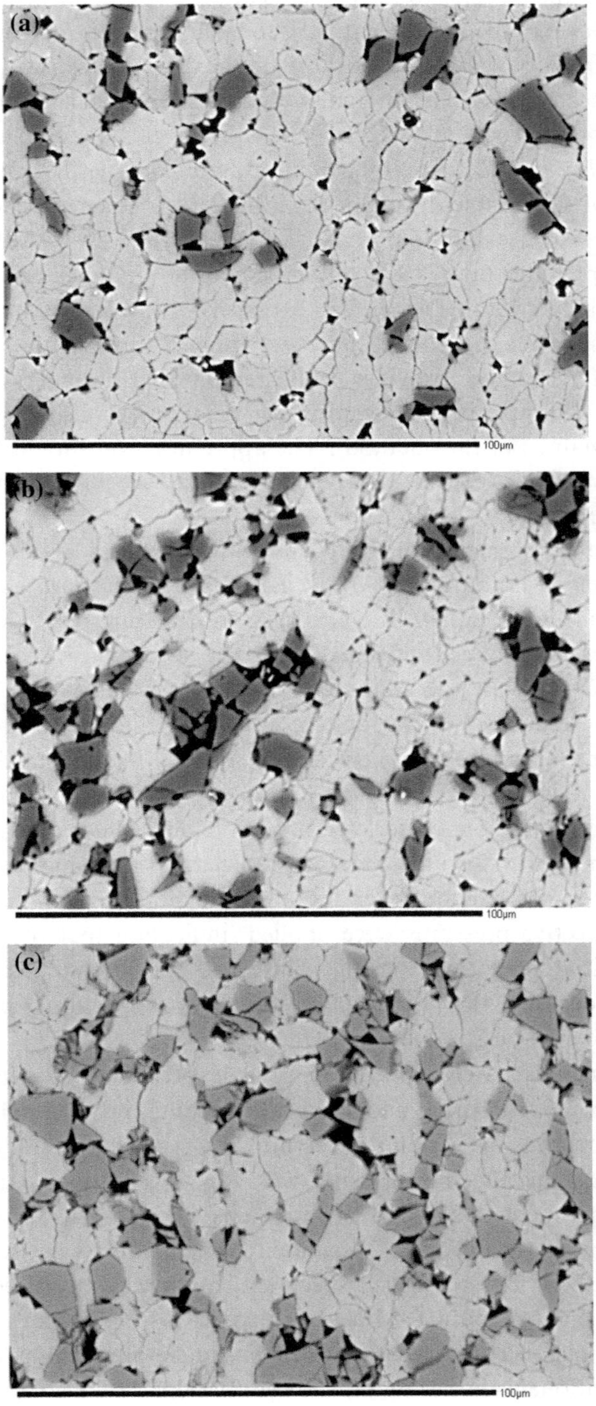
(a)
100μm
(b)
100μm
(c)
100μm

unconnected pore space variation is $dV_u = dV_{cu} - dV_{uc} + dV_{uu}$. In these relations, the subscripts *cc* and *uu* refer to the portions of pore space that remain connected or unconnected during the increment of compaction considered. Similarly, the subscripts *cu* and *uc* denote the portion of the pore space, originally connected that is disconnected, or that originally unconnected that becomes reconnected, respectively. Note that $dV_{uu}$ and $dV_{cc}$ may be either negative (compaction) or positive (dilation), but $dV_{uc}$ and $dV_{cu}$ are, by definition, $\leq 0$. Clearly, the volume of fluid $dV_f$ expelled from of the sample during *dt* (i.e., the volumometer measurement) is equal to $dV_{cc}$ and the decrease in total porosity is given by $d\phi = \frac{dV_{cc}+dV_{uu}}{V}$. If the pore fluid is not overly compressible (at 873 K, compressibility of argon is about 10 GPa$^{-1}$ at 50 MPa and 1 GPa$^{-1}$ at 290 MPa), we can expect $dV_{uu}$ to be very small since the pore pressure increase inside the isolated pores will prevent further pore shrinking. Thus, the volumometer method provides a (generally slight) underestimation of the variation of total sample porosity during compaction (see also ZHANG *et al.*, 1994, for an earlier analysis of this method). The approximation should be quite good at high porosities but is expected to degrade for porosities near or below the percolation threshold.

To start the experiment, confining and pore pressures of about 230 and 220 MPa, respectively, were applied. Argon gas was used as both confining pressure and pore fluid medium, nonetheless the pressures of each reservoir, ($\sigma_c$ and $P_{\text{fluid}}$, respectively) were maintained independently. The specimen was heated at a rate of 15 K/min to 873 K while the confining and pore pressures were adjusted to 300 and 290 MPa, respectively. The system was then allowed to settle under low effective pressure (about 10 MPa) for about an hour. After that, temperature and pore pressures were changed to the required values. This protocol provided uniform starting conditions before deformation and allowed us to estimate the pore space change with greater certainty. Indeed, the volumometer data showed very little further compaction after the run temperature had been reached, but before larger effective pressures were applied, indicating that the porosity change during this initial stage was small. Therefore, it is assumed that after the initialization procedure, the porosity was the same as the initial porosity at room temperature and pressure.

Pore pressure steps were applied in some experiments to determine porosity reduction and compaction rate under isostatic loading, i.e., the only external stress applied to the rock was the confining pressure, ($\sigma_1 = \sigma_2 = \sigma_3 = P_c$). In the isostatic compaction tests, the effective pressure was $P_{\text{eff}} = P_c - P_{\text{fluid}}$. In these isostatic compaction experiments, the total porosity change was also estimated from the change in length of the specimen, assuming compaction to be isotropic (see also ZHANG *et al.*, 1994). The change in specimen length was measured by applying a small load of 0.5 KN (about 6 MPa, less than 10% of the specimen strength) to ensure that the piston contacted the specimen, and noting the axial displacement reading. An additional calibration test showed that the axial displacement associated with the elastic compaction of the loading column during loading was negligible. We were able to measure displacements of the axial loading piston with a precision of $\pm 0.3$ μm, which, when converted to volumetric strain, corresponds to a resolution of $4.5 \times 10^{-5}$.

Finally, samples were deformed under conventional triaxial loading ($\sigma_1 > P_c$) at confining pressure of 300 MPa, pore pressures between 50 to 290 MPa, temperatures from 673 K to 1073 K and at strain rate of $3 \times 10^{-5}\ s^{-1}$. In the conventional triaxial tests, the effective mean stress was $P_{eff} = \frac{(\sigma_1+2\sigma_3)}{3} - P_{fluid}$. The load was recorded with an internally mounted load cell to a resolution of 2.5 N. In practice, hysteresis and drift in the zero of the load cell cause the axial load to be uncertain by 20 N, or about 0.3 MPa in axial stress. Displacements of the axial loading piston were recorded with a precision of $\pm 0.3$ μm, which is equivalent to a resolution of about $1.5 \times 10^{-5}$ in axial strain. During these tests, porosity variations could only be estimated from the volumometer recording. Important differences in behavior with respect to the isostatic tests are expected since compaction of the rock under triaxial loading is driven by both the effective pressure and the deviatoric stress resulting from the end load (see for example, BUDIANSKY *et al.*, 1982; WONG, 1990; and XIAO and EVANS, 2003).

### 2.3. *in-situ Permeability Measurements*

The oscillating pressure method was used to measure permeability continuously during both isostatic and conventional triaxial loading. At constant temperature and pore pressure, the servo-controlled motor drove the volumometer piston and generated a sinusoidal oscillation of fluid pressure in the upstream reservoir, while the resulting pressure variations in the downstream reservoir were recorded. The downstream signal must, in principle, be sinusoidal with the same frequency as the imposed upstream signal. In order to filter out the noise, Fourier analysis was used to calculate the ratio of downstream to upstream amplitudes, $\alpha$, and the phase shift between them, $\theta$. From these two parameters permeability and storage capacity can be inferred. If the pressure in the upstream reservoir is

$$P_{up} = P \sin(\omega t), \tag{1}$$

then the steady-state solution for pressure in the downstream reservoir is (FISCHER, 1992)

$$P_{down} = \alpha P \sin(\omega t - \theta),$$

$$\alpha e^{-i\theta} = \left( \frac{1+i}{\sqrt{\xi\eta}} \sinh\left[(1+i)\sqrt{\frac{\xi}{\eta}}\right] + \cosh\left[(1+i)\sqrt{\frac{\xi}{\eta}}\right]\right)^{-1}. \tag{2}$$

The dimensionless parameters $\xi$ (i.e., dimensionless storativity) and $\eta$ (i.e., dimensionless permeability) are defined by:

$$\xi = \frac{SL\beta}{\beta_D}, \quad \eta = \frac{STk}{\pi L \mu \beta_D}, \tag{3}$$

where $S$ is the sample cross-section area, $L$ the sample length, $\beta$ the sample storativity, $\beta_D$ the downstream reservoir storage, $k$ the sample permeability and $\mu$ the fluid viscosity (see also BERNABÉ *et al.*, 2006). Equation (2) was solved numerically to obtain the dimensionless parameters, $\xi$ and $\eta$, and then $k$ and $\beta$ were determined from equation (3). The amplitude of the pressure oscillations used in our experiments was 0.5 MPa, and the oscillation period ranged between 10 and 100 seconds. Note also that the oscillating pressure technique allows porosity variations to be inferred from the volumometer data time-averaged over one-period intervals.

The main source of uncertainties in the calculation of $k$ and $\beta$ is from the uncertainties on $\alpha$ and $\theta$ ($\delta\alpha$ and $\delta\theta$, respectively). In most cases, fast Fourier transforms yield $\delta\alpha/\alpha \approx 0.1\%-1\%$, and $\delta\theta/\theta \approx 2\%-5\%$, both of which, in general, are higher at higher frequency and elevated temperature. The uncertainties of $k$ and $\beta$ can be determined by generating random fluctuations of the input data ($\alpha + \delta\alpha$, $\theta + \delta\theta$), solving equation (2) and analyzing the statistics of the solutions (BERNABÉ *et al.*, 2006). In typical experiment conditions, the results show that uncertainties of $k$ are less than 10%, but, because the downstream storage is very large compared to the sample storage, $\beta$ can be underestimated by as much as 2 orders of magnitude.

## 3. *Results*

### *Isostatic Loading*

A comparison of porosity change determined from volumometer and from change in specimen length is shown in Figure 2 for an isostatic compaction experiment on a 20 wt% quartz sample at effective pressure of 200 MPa, The two porosity measurements agree well. In an experiment on 30 wt% quartz sample (Fig. 3a), confining pressure ($P_c$) was maintained constant at 300 MPa and pore pressure ($P_p$) was decreased step-wise from 250 to 100 MPa. Therefore, the effective pressure ($P_{\text{eff}} = P_c - P_p$) increased by steps of 50 MPa to 200 MPa. At each step, permeability measurements were made using the oscillating pressure method. The porosity changes shown in Figure 3b were inferred from the measured sample length variations. After each $P_{\text{eff}}$ increase, porosity drops rapidly at first, but a much slower compaction rate ($\dot{\phi}$) is eventually established. Relatively constant (i.e., very slowly decreasing) rates of permeability loss ($\dot{k}$) were also observed during these hold periods (Fig. 3b). Both $\dot{k}$ and $\dot{\phi}$ increase with increasing $P_{\text{eff}}$ and temperature. Figure 4 plots the logarithm of the relative permeability reduction rate $\dot{k}/k_0$ versus that of the relative compaction rate $\dot{\phi}/\phi_0$ for 10, 20 and 30 wt% quartz samples, where $k_0$ and $\phi_0$ are the starting permeability and porosity at each step, respectively.

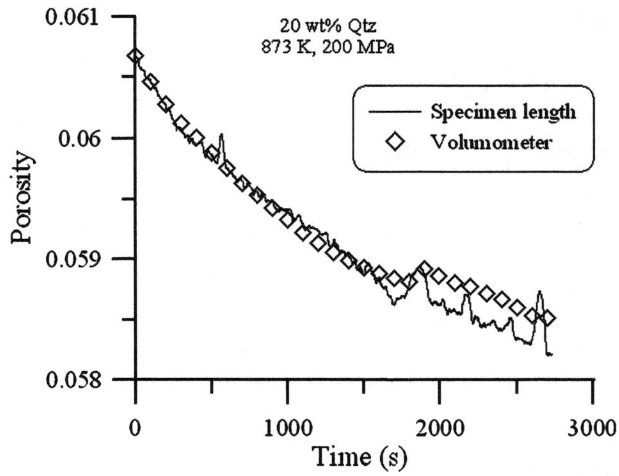


Figure 2
Isostatic compaction curves for a 20 wt% quartz sample subjected to effective pressure of 200 MPa at temperature of 873 K. The two porosity changes, determined from the volumometer and from the change in sample length show good agreement. Responses of axial actuator to the hysteresis and drift of the load reading sometimes cause fluctuation in the length record.

Rather than investigating the evolution of $\dot{k}$ and $\dot{\phi}$ during a single compaction period, our goal here is to show the relative difference of $\dot{k}$ and $\dot{\phi}$ for the samples with different quartz content at various effective pressures. Therefore, we argue that normalizing $\dot{k}$ and $\dot{\phi}$ with $k_0$ and $\phi_0$, respectively, can better serve this purpose. Over this permeability interval, the two logarithms are approximately linearly related, although there is enough scatter to obscure moderate nonlinearity (e.g., 30% data). The permeability reduction rate of 10 wt% quartz samples is usually higher than that for the 20 and 30 wt% samples in the high $\dot{\phi}/\phi_0$ regime. The relationship seems to reverse as $\dot{\phi}/\phi_0$ decreases, but the variation in the values for the different compositions is of the same order as the uncertainty in the data.

In Figure 5, permeability is plotted against porosity under isostatic loading at constant temperature of 873 K with effective pressure increasing by steps from 10 MPa to 250 MPa. The data can be described using local fits to a power-law, $k \propto \phi^n$, with $n$ increasing with decreasing porosity, until the total porosity approaches a finite value, at which the permeability is too small to be measured accurately (less than $10^{-21}$ $m^2$). This "cutoff" porosity for the samples with 10 wt% quartz appeared to be lower than that for the 20 wt% and 30 wt% samples, although the isostatic compaction experiments were of insufficiently long duration for the latter samples to effectively reach the cutoff porosities.

Storativity ($\beta$) measurements, as already mentioned, are subject to greater uncertainty than permeability data (Fig. 6), owing to the large downstream reservoir storage. Nevertheless, we observed a general trend of decreasing $\beta$ with decreasing

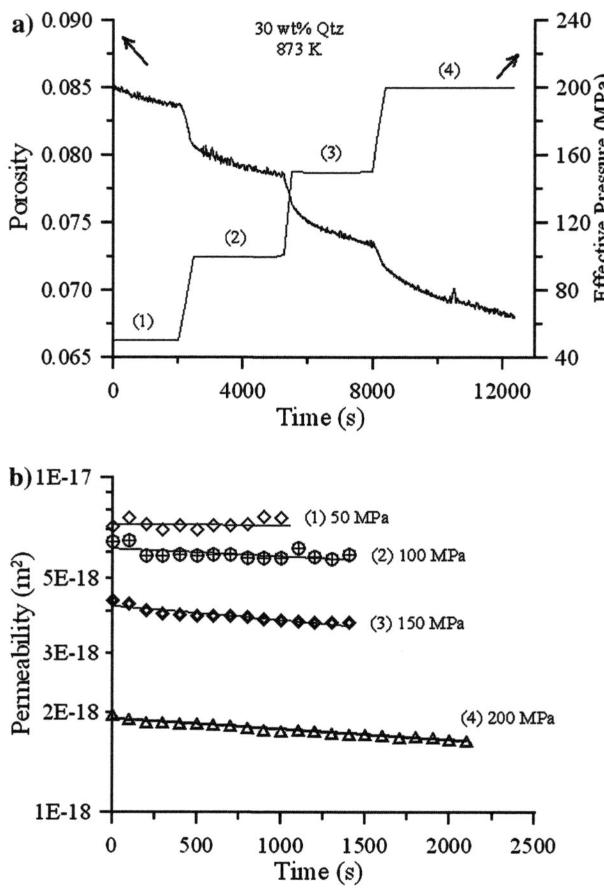


Figure 3

Evolution of (a) porosity and (b) permeability as a function of time for a 30 wt% quartz sample compacted under isostatic loading. Several different effective pressure steps (1–4) were applied to this sample; the effective pressure history is indicated in the right hand ordinate. In (b), the permeability evolution is given as a function of the elapsed time from the application of the appropriate pressure for each step, i.e., $t = 0$ is reset for each step.

porosity for 30 wt% quartz samples. However for the 10 and 20 wt% quartz sample, $\beta$ remained roughly constant.

In some experiments permeability measurements were made during additional cycles of unloading and loading (i.e., cycles of increasing and decreasing $P_{eff}$). The data show that there was significant non-recoverable reduction in permeability between the first loading and unloading (Fig. 7). Changes of permeability during subsequent effective pressure cycling were recoverable and less significant, probably because of the low compaction rate and relatively shorter holding time at each subsequent pressure step. Therefore, data obtained for later pressure cycles probably

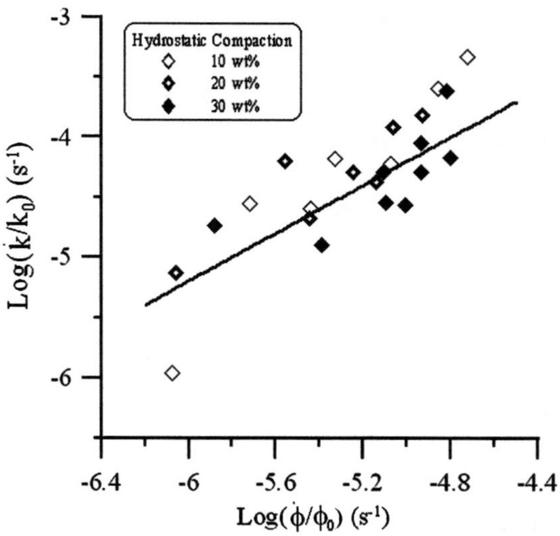


Figure 4
Relation between relative permeability reduction rate and relative porosity for isostatic compaction experiments. Permeability reduction rate $\dot{k}/k_0$ versus compaction rate $\dot{\phi}/\phi_0$ for 10, 20 and 30 wt% quartz samples, where $k_0$ and $\phi_0$ are respectively the starting permeability and porosity at each step.

represents the permeability dependence on effective pressure associated with elastic deformation (Fig. 8a). On log scale, the pressure sensitivity of permeability appears similar for all 10 wt%, 20 wt% and 30 wt% samples.

Figure 8b shows storativity $\beta$ measured at different effective pressures. Although there is considerable scatter, it can be seen that storativity of 30 wt% and 20 wt% samples decreases with increasing effective pressure, while 10 wt% samples show very little pressure sensitivity. In most cases, storativity seems to increase with increasing quartz content, although the data are quite scattered.

*Conventional Triaxial Loading*

Under triaxial loading, the samples compacted as they deformed, except during the tests run at an effective pressure of 10 MPa, where porosity and permeability remained approximately constant. At high effective pressure (>200 MPa) and particularly for 10 wt% quartz samples, *in situ* permeability measurements were limited because very low values of permeability required long periods (more than 200 s). With the axial strain rate at $3 \times 10^{-5}$ s$^{-1}$, the porosity change was substantial even over the time required for a single oscillation. In general, during the triaxial experiments, the differential stress, porosity, permeability and storativity all changed as a function of axial strain. Typical triaxial deformation experiments conducted at an effective pressure of 100 MPa are shown in Figure 9. For all 10, 20 and 30 wt% samples, the mechanical data exhibited an accelerated compaction in comparison to

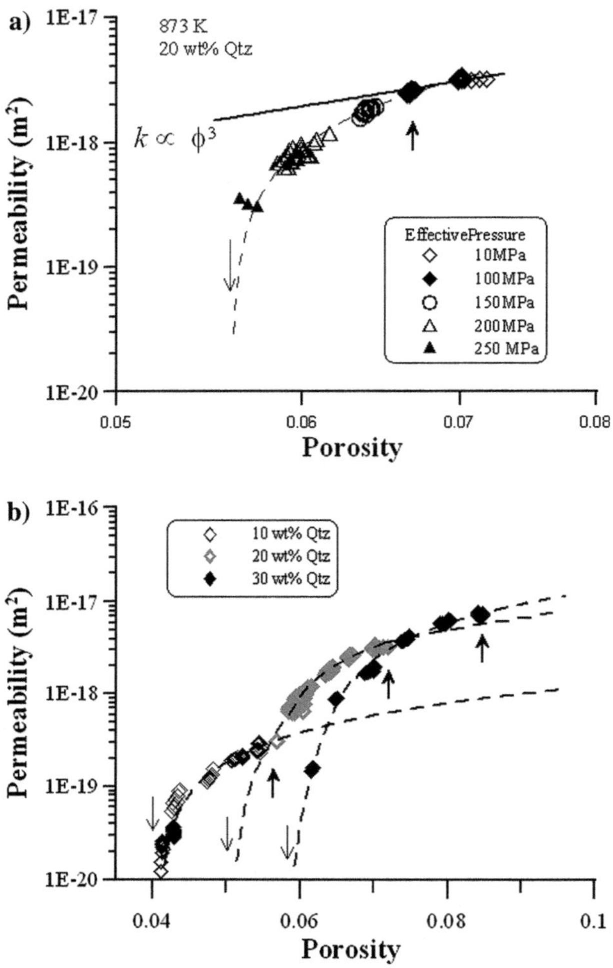


Figure 5
Permeability evolution as a function of porosity under isostatic loading at constant temperature of 873 K with effective pressure increasing by steps from 10 MPa to 250 MPa for 20 wt% quartz samples (a). The convex, non-linear relationship between $k$ and $\phi$ is often observed in materials compacting by viscous creep under isotropic loads (BERNABÉ *et al.*, 2003). If the relationship is expressed locally as $k \propto \phi^n$, then the exponent $n$ increases with decreasing porosity. The porosity at which the curve departs substantially from $n = 3$ is called the crossover porosity; the porosity at which the permeability becomes too low to measure is a measure of a percolation threshold. (b) Comparison of the data for 10, 20 and 30 wt% quartz samples, both the cross-over porosity (thick upward arrows) and the threshold porosity (thin downward arrows) seem to depend inversely on the quartz content.

compaction rates measured during isostatic tests. We also observe (e.g., Fig. 9) that permeability in the triaxially loaded samples closely tracked the decreasing porosity. On the other hand, storativity appeared to be roughly constant during deformation, but the reduced accuracy of this measurement must be kept in mind.

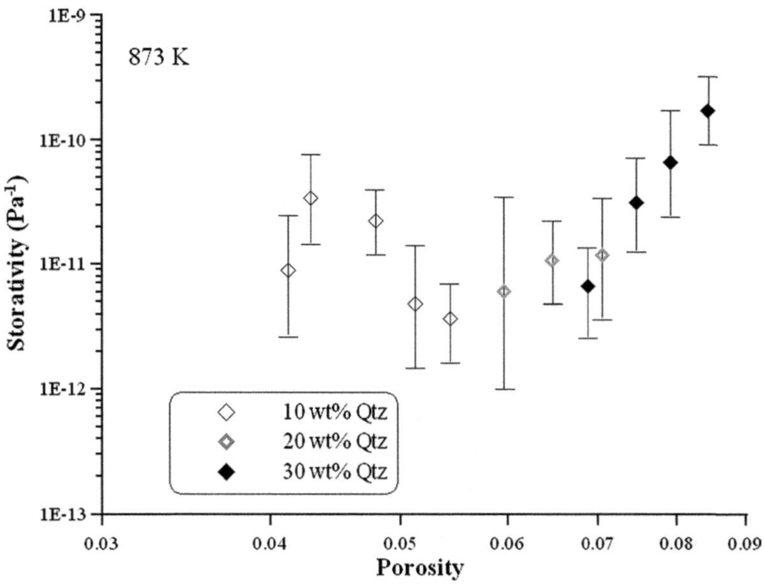


Figure 6
Evolution of storativity of the aggregates under isostatic loading. Storativity (sometimes called the specific storage) is $\beta = \frac{1}{V_{agg}} \cdot \frac{dV_{fluid}}{dP_{fluid}}$. In general $\beta$ decreases with decreasing porosity for 20 and 30 wt% quartz samples, but for the 10 wt% quartz samples, it remains roughly constant. Owing to constraints of the apparatus, the determination of storativity is less accurate than that of the permeability. The error bars represent $\pm 2$ standard deviations.

Since the evolution of permeability and porosity in rocks is dictated by the mechanisms of change of pore space (BERNABÉ *et al.*, 2003), systematic characterization of the evolution curves may help to provide clues to the operating process. We focus on the effect of quartz fraction, pressure and temperature on the exponent $n$, assuming a piece-wise power-law description, $k \propto \phi^n$ with a variable $n$.

*Quartz content effects.* In Figure 10, permeability is plotted as a function of porosity for 10, 20 and 30 wt% samples deformed at 873 K and various effective pressures from 50 MPa to 250 MPa. At effective pressure of 100 MPa, the exponent, $n$, increases from 1.6 for 30 wt% sample to 3.1 for 10 wt% sample.

*Pressure effects.* For each test the starting permeability and porosity varied, depending on the previous loading history. For example, sample 20g08 (20 wt% quartz) had previously compacted under an isostatic pressure of 250 MPa, and had undergone an additional deformation step at effective pressure of 200 MPa during which a significantly higher $n$ ($\sim$6.4) was observed; thus, that sample had a lower "starting" porosity than sample 20g10 (Fig. 10b). The data, together with those of 30 wt% quartz samples (Fig. 10c), show that increased effective pressure often results in higher $n$.

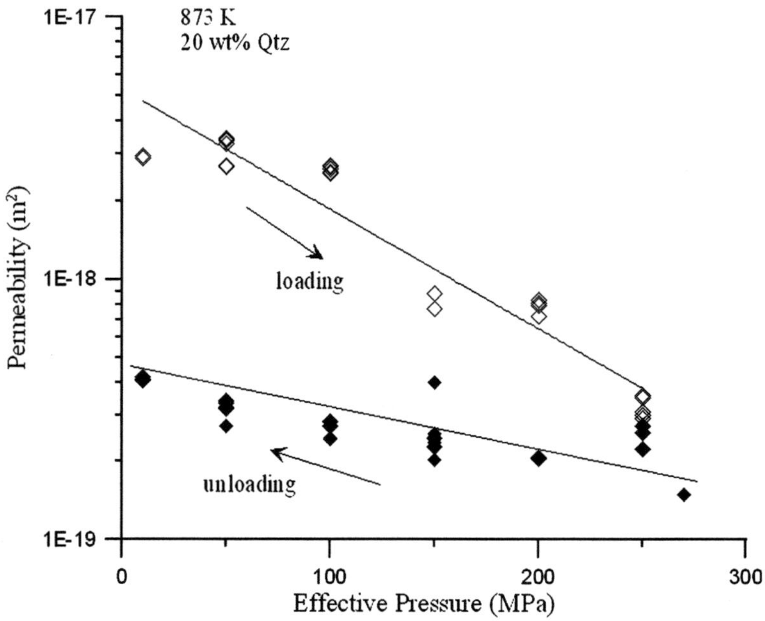


Figure 7
Permeability as a function of effective pressure during isostatic loading of a 20 wt% sample at 873 K. The sample was loaded in steps of increasing effective pressure; during each step, several measurements of permeability were performed. After the maximum effective pressure step (250 MPa) was attained, a sequence of measurements was done for steps at decreasing effective pressures. Altogether two cycles of increasing and decreasing effective pressures were generated. During the first loading cycle there was significant non-recoverable compaction, with an associated permeability loss. Because the compaction rates were considerably reduced on subsequent cycles, the permeability loss with increasing effective pressure probably measures elastic changes in the pore space. Presumably, larger elapsed times could result in inelastic compaction that would cause a non-recoverable permeability loss.

*Temperature effects.* A few more experiments at different temperatures were carried out for 20 wt% and 30 wt% quartz samples (Fig. 11). It seems that varying temperature from 773 K to 1073 K makes little difference in the $n$ values. For most 20 wt% samples (Fig. 11a), $n$ is 2.5–3.0 except for sample 20g08b, which, as mentioned above, was deformed previously at an effective pressure of 100 MPa. Small $n$ ($\sim$1.5) is generally found for 30 wt% samples, often following a rather large permeability drop which corresponds to relative small porosity reduction (or large $n \sim$10.0). This gradual transition of $n$ values usually occurred at porosities around 7.0%.

In summary, the most striking result is that the permeability decrease for a given increment of compaction is smaller under triaxial loading than for isostatic loading (Fig. 10). During isostatic loading, permeability reductions for a given incremental decrease in porosity are increasingly larger as porosity decreases, and therefore the exponent $n$ increases from 3 to 20. Paradoxically, while the time rate of change of

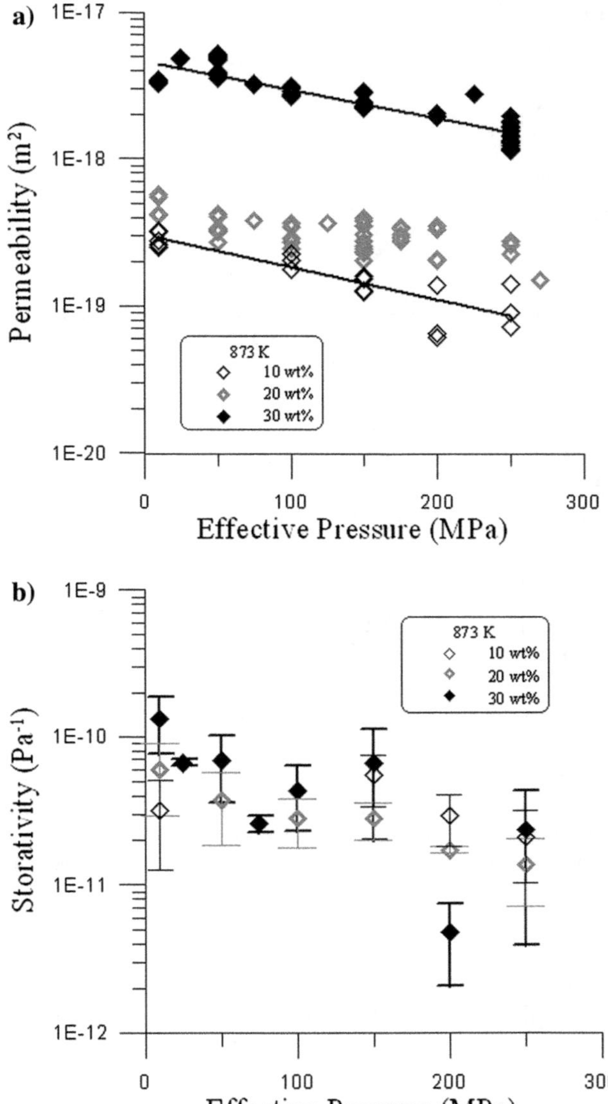


Figure 8
Permeability and storativity of samples subjected to a series of increasing and decreasing isostatic loads. Permeability (a) and storativity (b) are plotted as function of effective pressure for 10, 20 and 30 wt% quartz samples. The properties plotted here are those measured after the first cycle of increasing pressure steps (as in Fig. 7 above). For an elastic porous material, the slope of the curves is a qualitative measure of the shape of the pore space.

porosity is larger under triaxial loading than under isostatic loading, permeability reductions are smaller for a given incremental decrease of porosity. Consequently, the permeability-porosity curves of triaxial deformation depart sharply from that

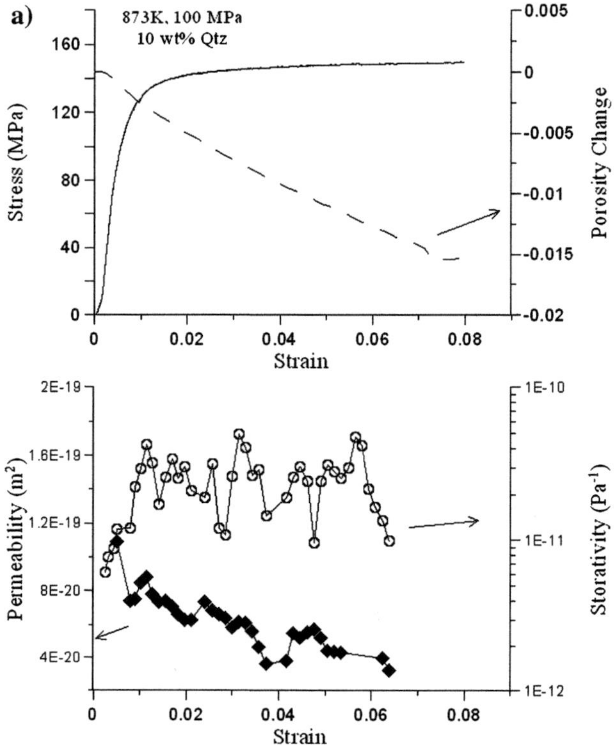


Figure 9
Results for samples loaded in conventional triaxial configuration at 873 K, effective pressure ($\sigma_3 - P_f$) of 100 MPa. Upper: Differential stress (solid line) and porosity change (dash line) versus axial strain $\varepsilon_1 = \ln(l - l_o/l_o)$; Lower: Permeability (diamond symbol) and storativity (circle symbol) for (a) 10 wt% (b) 20 wt%, and (c) 30 wt% quartz samples.

corresponding to isostatic compaction. For a given porosity, permeability under triaxial loading is higher than that developed during compaction under isostatic loading.

## 4. *Discussion and Implications*

In this study we investigated the permeability evolution of porous quartz-calcite aggregates as they underwent isostatic compaction and compaction during triaxial deformation. In both cases conditions were such that the local deformation mechanisms are probably dominated by dislocation creep (De Bresser and Spiers, 1993; Renner *et al.*, 2002; Rutter, 1974). Immediately below we compare the isostatic compaction data with previous studies on hot-pressed aggregates of monomineralic calcite to reveal some details of the effect of adding a second-phase.

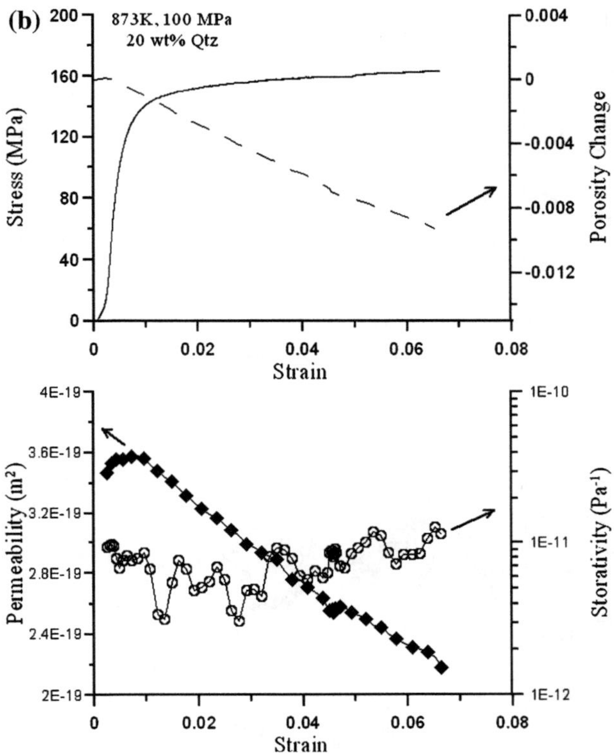


Figure 9
(Contd.)

Following that, we discuss the effect of loading conditions on the evolution of permeability, and, finally, consider some implications of the results for fluid migration, melt extraction and shear localization under natural conditions.

### *4.1. Isostatic Compaction*

#### *Comparison to Previous Data*

Several studies have measured the relationship between porosity and permeability in porous calcite aggregates under isostatic loading at elevated temperatures and pressure (BERNABÉ *et al.*, 1982; WARK and WATSON, 1998; ZHANG *et al.*, 1994b, 2000, 2001; ZHU *et al.*, 1999) (see Fig. 12). Differences in the data sets exist owing to variations in confining pressure, pore fluid chemistry, temperature, and testing techniques. For example, the high permeability measured by (WARK and WATSON, 1998) is likely due to the fact that those authors measured gas permeability under very low confining pressure after synthesis by hot isostatic pressing. In the other data suites, permeability was measured at elevated confining pressures, either after HIP

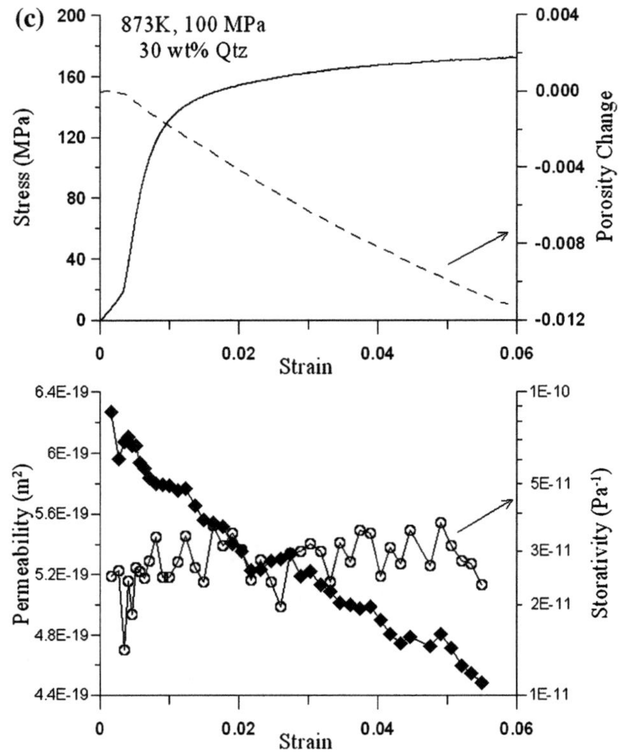


Figure 9
(Contd.)

(Bernabé *et al.,* 1982) or during HIP (Zhang *et al.,* 1994b, this study). Notwithstanding these differences, some generalizations can be made. First, the permeability reduction rate covaries with the compaction rate and increases with increasing effective pressure. Second, the permeability is nonlinearly related to porosity. Fits can be made over a portion of the data range using a piece-wise power-law relation: $k \propto \phi^n$. In most cases, the exponent, $n$, progressively increases as porosity decreases, and a finite cutoff porosity is reached at which the permeability becomes too low to be measured (Bernabé *et al.,* 1982; Zhang *et al.,* 1994b), indicating the existence of a porosity percolation threshold.

▶

Figure 10
Permeability versus porosity for (a) 10, (b) 20 and (c) 30 wt% samples deformed under triaxial loads at various effective pressures at 873 K. The lighter dashed curves are the same relations for samples deformed under isostatic loads. In general the samples deforming under triaxial loads have higher permeability (at least at low porosity) than their isostatically loaded counterparts. In addition, the permeability of the triaxially loaded samples is less sensitive to porosity decrease; and the slope of the curves ($n$) increases with increasing effective pressure.

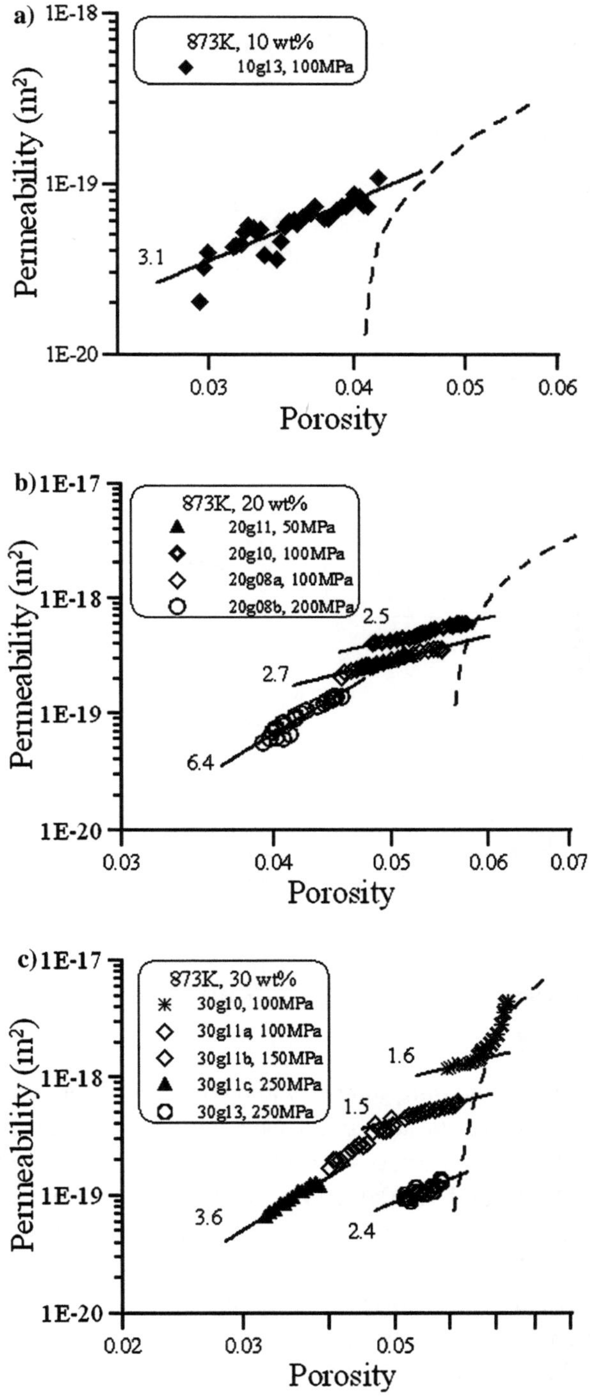
a)
873K, 10 wt%
10g13, 100MPa
3.1
Permeability (m²)
1E-18
1E-19
1E-20
0.03
0.04
0.05
0.06
Porosity
b)
873K, 20 wt%
20g11, 50MPa
20g10, 100MPa
20g08a, 100MPa
20g08b, 200MPa
2.5
2.7
6.4
Permeability (m²)
1E-17
1E-18
1E-19
1E-20
0.03
0.04
0.05
0.06
0.07
Porosity
c)
873K, 30 wt%
30g10, 100MPa
30g11a, 100MPa
30g11b, 150MPa
30g11c, 250MPa
30g13, 250MPa
1.6
1.5
3.6
2.4
Permeability (m²)
1E-17
1E-18
1E-19
1E-20
0.02
0.03
0.05
Porosity

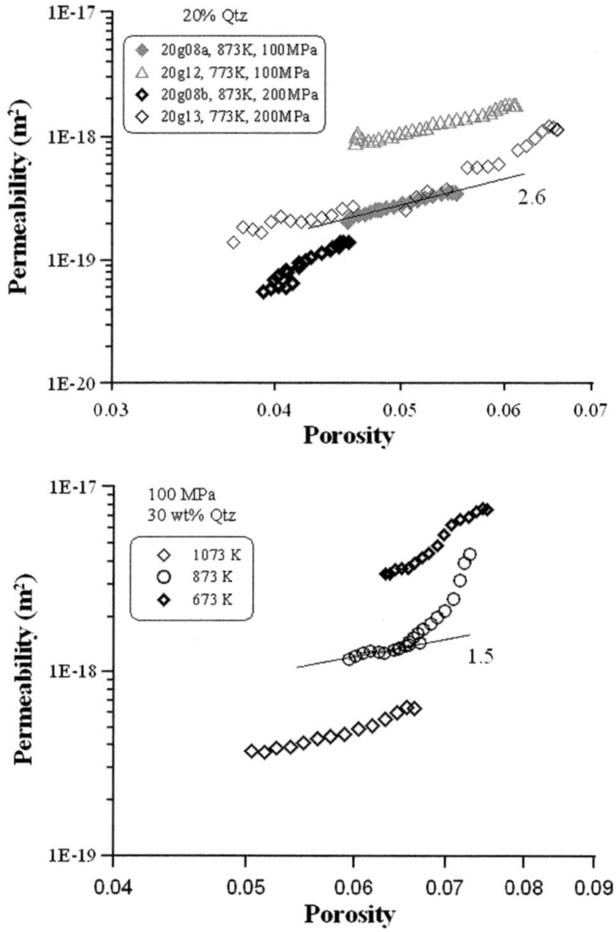


Figure 11
Permeability versus porosity for (a) 20 and (b) 30 wt% samples deformed at different temperatures.

It is well known that the pore geometry and connectivity of fluid or melt-bearing, texturally equilibrated rocks are controlled by the interfacial energies along solid-solid and solid-fluid interfaces (HOLNESS and GRAHAM, 1995; VAN BARGEN and WAFF, 1986; WARK and WATSON, 1998; WATSON and BRENAN, 1987). Thus, the permeability-porosity relationship can be predicted in general terms, based on the knowledge of dihedral angle, but, at present, complications from non-uniform grain size, multiple phases and anisotropy of surface energy cannot be easily incorporated. In these experiments, inert argon gas was used as the pore fluid. Under similar experimental conditions, equilibrated textures with dihedral angle around 77° have been observed (ZHANG *et al.*, 1994b), and thus, one might expect that a lower percolation porosity would occur when the pore fluid would support a lower dihedral angle (e.g., when the fluid is aqueous, HOLNESS and GRAHAM, 1995; WARK and WATSON, 1998).

*Percolation Porosity and Quartz Content*

Adding rigid second-phase particles certainly induces new complexity to the pore structure (Fig. 1), and, as expected, the changed pore structure affects the permeability (see BERNABÉ *et al.*, 2004). The permeability reduction rate is also influenced: Compaction rates in 10 wt% quartz samples are generally slightly higher than those in the 30 wt% samples at the same total porosity and effective pressure.

The exact effect of the second phase on the pore structure of the aggregate probably cannot be predicted from theory without a much more detailed description of the mineral distribution. SEM photomicrographs indicate that clusters of quartz grains exist. These aggregates were probably formed during mixing and then incorporated into the matrix during HIP. For spatial scales larger than the dimensions of the clumps, the phase distribution is relatively homogeneous. As argued by BERNABÉ *et al.* (2004), the calcite-quartz rocks can then be described as a binary mixture of materials of high-permeability (quartz cluster) and low-permeability (calcite matrix).

In Figure 5b we compile the data for permeability as a function of porosity during isostatic compaction. All samples exhibit similar permeability-porosity

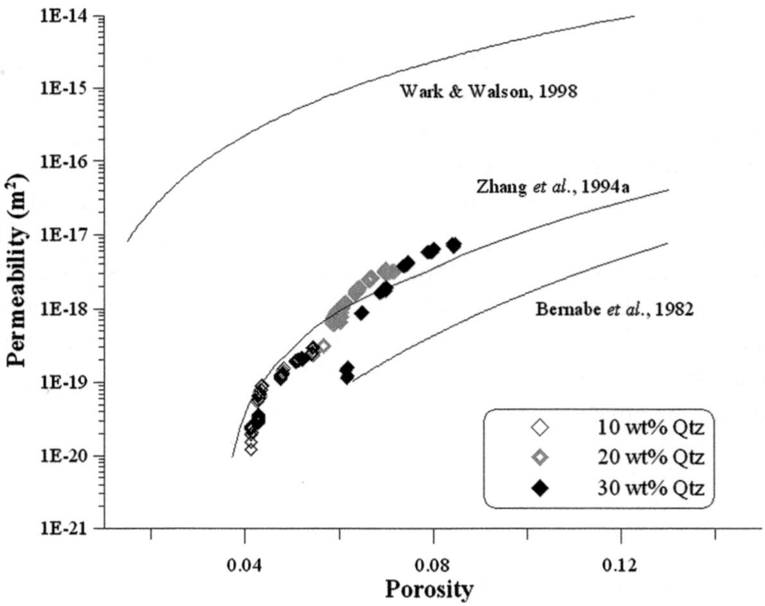


Figure 12
Isostatic compaction data in comparison with previous studies on hot-pressured calcite aggregates (BERNABÉ *et al.,* 1982; WARK and WATSON, 1998; ZHANG *et al.,* 1994b).

relationships, i.e., the sensitivity of permeability to porosity change increases as porosity decreases. At large porosity, the permeability/porosity relationship is a power law with $n$ very close to 3, but $n$ increases with decreasing porosity. As the material compacts, total porosity asymptotically approaches a percolation (or cutoff) value below which pore connectivity is completely lost (BERNABÉ *et al.,* 1982; WARK and WATSON, 1998; ZHANG *et al.,* 1994b). Good agreement with ZHANG *et al.* (1994b) is expected (Fig. 12), because of the similar experimental conditions and identical experimental techniques.

Isostatic compaction curves of this form have been interpreted as the final stage of HIP, where tube ovulation disconnects small pore throats connecting larger pores at multi-grain junctions (EVANS *et al.,* 1999; ZHU *et al.,* 1999; HICKMAN and EVANS, 1987). The permeability-porosity relationship can be roughly characterized by the percolation threshold and the point at which the curve deviates substantially from $n = 3$, which, for convenience is called the crossover porosity (ZHU *et al.,* 1999). Following this conceptual model, we estimate the percolation and crossover porosity by extrapolating the experiment data to low and high porosity regime, respectively. Our data suggest that both percolation threshold and crossover porosity are positively correlated with quartz content, and that the permeability differences among the samples with different quartz fraction evolve with change of porosity (Fig. 5b). Namely, at high porosity, where all samples follow cubic law, permeability increases with quartz content. However, owing to the fact that both the crossover and percolation porosity occur at higher porosity for higher quartz contents, the 30 wt% quartz samples become less permeable than 20 wt% samples below crossover.

This behavior may arise from the differences in either ovulation rate or compaction kinetics. Owing to higher strength of quartz, the quartz clusters likely compact less and maintain high permeability, but the surrounding calcite matrix may be compacting faster, or tube ovulation may be occurring more quickly than in samples with lower quartz content. Increased compaction rates in the calcite matrix might result from stress risers around the strong quartz clusters. The ovulation rate might covary with quartz content, because, for a given total porosity, the dimensions of the pore throats in the calcite matrix might be smaller if a larger proportion of the porosity were partitioned into the quartz clusters. The increased percolation porosity might then be associated with the unconnected porosity stranded in the quartz clusters.

To make this point clearer, consider the following, often used, effective medium model:

$$k = \frac{d^2 \phi^m}{C}, \tag{4}$$

where $d$ is the grain size, and, $C$ and $m$ are constants. During compaction, as the pores along grain boundaries were progressively disconnected, impermeable domains of calcite matrix were formed. Consequently, $d$ in Equation (4) increased to the diameter of the impermeable domain, which is limited by the mean distance between quartz clusters. Obviously, the space between quartz clusters must be inversely proportional to quartz fraction. Therefore, samples with equal porosity but higher quartz content are expected to have lower permeability. In the case of the partially molten peridotite DAINES and KOHLSTEDT (1993) suggested that increase of melt-free triple junctions due to the presence of pyroxene could result in an increase of the size of melt channels if melt fraction remains constant, leading to an enhanced melt migration rate. WARK and WATSON (1998) proposed a similar idea that closure of some grain-edge channels due to distribution of dihedral angles would increase the "effective grain diameter" and increase the bulk permeability. Although these arguments are qualitatively useful, to predict rates of compaction accurately will require full consideration of the heterogeneities of stress, porosity, and mechanical properties within the aggregate.

### *4.2. The Effect of Viscous Creep on Pore Geometry*

When subjected to triaxial loading, the calcite-quartz aggregates exhibit shear-enhanced compaction, although, counter to intuition, permeability is not as sensitive to changes in porosity as under isostatic loading (Fig. 10). The exponent $n$ (from $k \propto \phi^n$) varies from 2 to 3, even at fairly low porosity. Hence, the triaxially loaded samples remain permeable even when the porosity drops below the percolation threshold for the samples compacted under isostatic loads. In other words, non-isostatic deformation reduces the percolation porosity, and, in fact, enhances the permeability relative to that for the same porosity achieved under isostatic compaction.

Notice that the porosity data determined during triaxial deformation is determined using the volumometer. As stated in Section 2.2, this method approximately monitors changes in total porosity although it yields an underestimation of the total compaction rate. During triaxial tests, some unconnected pores, inside which the local fluid pressure must be higher than the imposed pore pressure owing to compaction, may reconnect to the connected pore space. Repartitioning of the closed porosity may, in fact, be one process contributing to permeability retention. The difference in the permeability-porosity relationships of samples compacted under different loading conditions highlights, once again, the fact that permeability is a function of all aspects of the pore microstructure, including such parameters as connectivity, pore size distribution, network geometry, rather than just the value of the total porosity.

#### *Connectivity*

It is generally believed that the departure from a simple (between square and cubic) power law of the permeability-porosity relationship (i.e., increase of exponent $n$) produced by compaction is probably related to the loss of pore connectivity (BERNABÉ

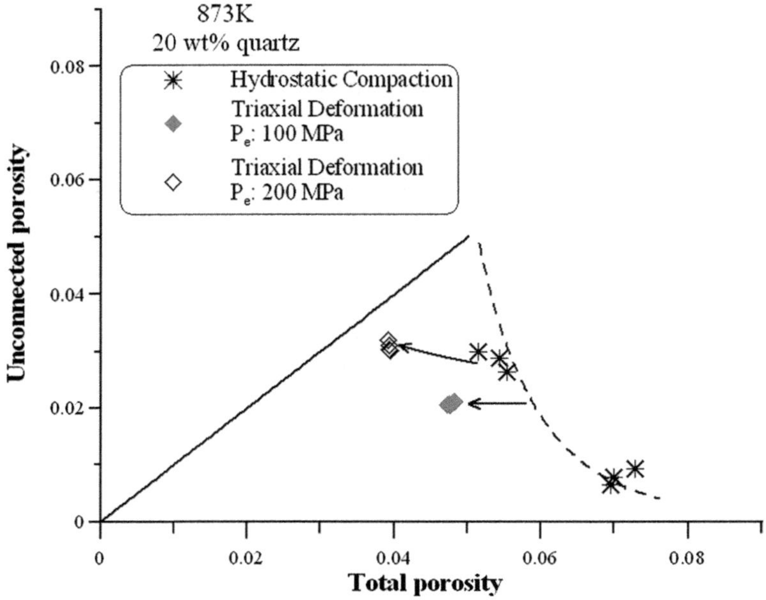


Figure 13
(a) Total porosity versus unconnected porosity for 20 wt% quartz sample under isostatic compaction and triaxial deformation. The dashed line is the predicted curve obtained by extrapolating the permeability-porosity curve from above the crossover porosity to a value below which we cannot measure permeability. Arrows indicate the porosities before and after deformation under triaxial loading.

*et al.*, 1982, 2003; ZHANG *et al.*, 1994b). Thus, as a first-order approximation, the porosity at which the departure occurs (crossover porosity) can be used to identify the onset of rapid disconnection (ZHU *et al.*, 1999). Thus, for the samples loaded triaxially, the observation that permeability is roughly proportional to a power between 1.5 and 3.5 of porosity implies that pore connectivity is relatively constant during triaxial deformation. Measurements of the unconnected porosity made on the run products also indicate that the triaxially deformed samples have a smaller proportion of unconnected porosity than those compacted under isostatic loading (Fig. 13).

*Aspect Ratio*

In addition to repartitioning of open and closed porosity, high connectivity could be maintained during deformation by changing the pore shape in a way that would increase transmissivity. Using the theory of elasticity of porous bodies (WALSH, 1965), an average or representative pore aspect ratio ($a/c$) can be estimated as (FISCHER and PATERSON, 1992):

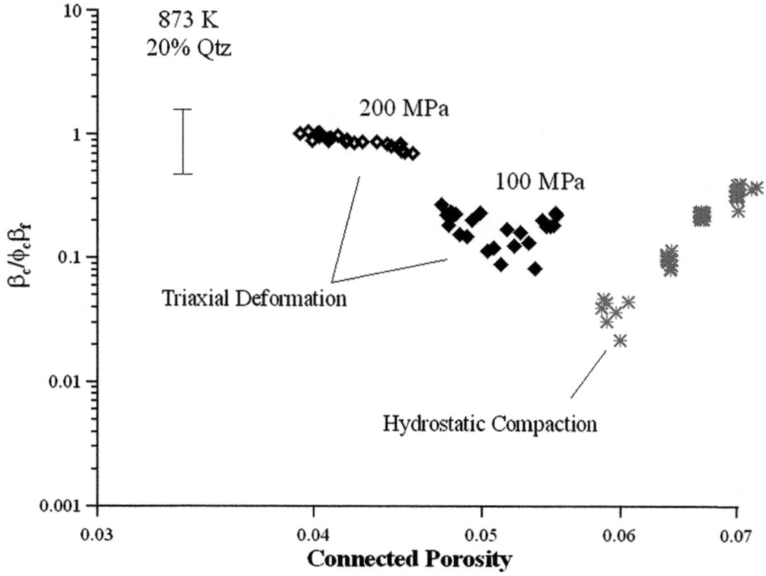


Figure 14
Evolution of storativity with decreasing porosity for 20 wt% quartz samples deformed in triaxial loading with that during isostatic compaction.

$$\frac{a}{c} \approx 1 + \frac{\beta_f}{\beta_m}\left(\frac{\beta_c}{\phi_c\beta_f} - 1\right), \tag{5}$$

where $a$ and $c$ are the maximum and minimum dimensions of the pore respectively, $\beta_m$ and $\beta_f$ are compressibilities of the mineral and the pore fluid, respectively, and $\beta_c$ and $\phi_c$ are the storativity and connected porosity of the sample, respectively. The values of $\beta_c/(\phi_c\beta_f)$ are plotted against porosity of, for example, 20 wt% quartz samples in Figure 14. The connected porosities $\phi_c$, under isostatic loading, are inferred from extrapolating permeability-porosity (total) curves from above cross-over porosity to cut-off porosity, assuming a cubic power law $k \propto \phi_c^3$. Using the $\phi_c$ measured before deformation, and the observation that there is little change of unconnected porosity during deformation (Fig. 13), $\phi_c$ under triaxial loading can be estimated. We see that values of $\beta_c/(\phi_c\beta_f)$ are mostly less than 1, which is problematic, because the storativity of fluid/rock system should not be less than that of fluid alone, implying that $\beta_c/(\phi_c\beta_f)$ cannot be less than unity (note that FISCHER and PATERSON (1992) encountered the same problem). The reason for these discrepencies is probably that $\beta_c$ was significantly underestimated here (see Section 2.3). However, we may still assume that the trends displayed in Figure 14 are meaningful. Then, the data suggest that $\beta_c/(\phi_c\beta_f)$, and consequently the aspect ratio $a/c$, increases as porosity decreases, indicating that the pores become less equant. Indeed, observations of microstructure did show that many spherical pores in calcite matrix were eliminated by plastic deformation (XIAO and EVANS, 2003). Furthermore, under axisymmetric stressing, an initially spherical void in nonlinearly viscous media tends to evolve towards an oblate shape (BUDIANSKY *et al.*, 1982). Therefore, it is not unreasonable to conclude that the pore aspect ratio $a/c$ did increase during triaxial deformation.

The percolation threshold $\phi_p$ is also known to be related to aspect ratio. For example, in the case of a population of penny-shaped cracks homogeneously distributed throughout the rock, the percolation threshold is given by (GUÉGUEN and DIENES, 1989):

$$\phi_p = \frac{4}{3\pi}\frac{c}{a}. \tag{6}$$

Thus, increases in $a/c$ should also result in decreases of the percolation threshold, which is consistent with our experimental observations.

### *4.3. Fluid Flow in Deformable Porous Rocks*

The fundamental problems concerning fluid flow in deformable porous medium are pertinent to melt extraction from partially molten rocks, expulsion of sedimentary fluids in compacting rocks, and to fluid flow in crustal rocks undergoing tectonic deformation or metamorphic reactions. It is commonly recognized that a system of two intermingled viscous fluids (e.g., a deformable solid matrix saturated with fluid) is unstable with respect to grain scale distribution of fluid, if the presence of the fluid introduces gradients in chemical potential (CONNOLLY, 1997; MCKENZIE, 1984; SCOTT, 1988; STEVENSON, 1989). The underlying assumption is that the shear viscosity of the aggregate depends on the fluid fraction, which means lower mean pressure in the regions with high fluid fraction if the strain rate is constant. Pressure gradient drives fluid into the fluid-rich region, causing further lower pressure. Thus, instability or fluid segregation forms over a scale called the compaction length $\delta_c$ (MCKENZIE, 1984):

$$\delta_c = \left[\frac{k(\lambda + 4/3\eta)}{\mu}\right]^{1/2}, \tag{7}$$

where $\lambda$, and $\eta$ are the bulk and shear viscosity of the aggregate, respectively, and $\mu$ is the liquid viscosity. Since we directly measured permeability and compaction rate of the samples under deformation, the experimental data constrain the aggregate parameters. To obtain an estimate of the compaction length using this formulation, we computed an average value for the bulk and shear viscosities of the calcite-quartz aggregates. In so doing, we disregarded the effect on strength of variations of quartz content and porosities (for which, see XIAO and EVANS, 2003).

Based on the isostatic compaction data, the bulk viscosity, $\lambda$, is about $10^{14}$–$10^{15}$ Pa s at 873K and an effective pressure of 100 MPa, following $\lambda = P_e/\dot{\varepsilon}_v$, where $P_e$ is the effective pressure and $\dot{\varepsilon}_v$ is the compaction rate. At the same condition, the triaxial deformation data (flow strength of 150 MPa at strain rate of $3.0 \times 10^{-5}$) gives a value of $10^{11}$–$10^{12}$ Pa s for shear viscosity $\eta$, implying that $\lambda \gg \eta$. There are widely varying assumptions about the value of the term $(\lambda + \frac{4}{3}\eta)$, which is also referred as compaction viscosity of the matrix (SCOTT, 1988). For example, it is sometimes assumed that $\lambda \ll \eta$ for partially molten mantle rocks (HOLTZMAN *et al.*, 2003; RENNER *et al.*, 2003) and metamorphic rocks undergoing pressure solution (MCKENZIE and HOLNESS, 2000). In contrast, CONNOLLY (1997) suggests that for diffusion creep at constant grain size, $\lambda = \frac{3}{4}(\eta/\phi)$, based on the mechanical model of densification for HIP (WILKINSON and ASHBY, 1975). Similarly, a relationship, $\lambda = \frac{3}{4}(\eta/\phi^{1/2})$, is obtained for pressure solution (HELLE *et al.*, 1985). In fact, $\lambda$ and $\eta$ are defined in a way similar to that for Lamé's constants of elasticity. Following the model for granular flow (PATERSON, 1995) assisted by diffusion via an intergranular fluid, RUTTER (1997) suggests that $\lambda \approx 8\eta$ when pore pressure equals confining pressure. Thus, it is conceivable that the compaction viscosity $(\lambda + \frac{4}{3}\eta)$ is of the same order of magnitude as the bulk viscosity.

With values of $10^{-19}$ – $10^{-18}$m$^2$ for the permeability, $10^{14}$–$10^{15}$ Pa s for the bulk viscosity and $7 \times 10^{-5}$ Pa s for the viscosity of Argon gas, Equation (7) gives a compaction length of 0.1–1 m, which is about 10–100 times the sample length. If this calculation is pertinent, then fluid segregation would not be expected to occur at the sample scale, and no significant pore pressure gradient should develop. We thereby can rule out the possibilities of channeling or focused flow owing to spatial heterogeneities at a scale smaller than the sample in these experiments.

Clearly, in these experiments, the viscosities, $\lambda$, $\eta$, and permeability, $k$, of the calcite quartz rocks are dependent on porosity (see also XIAO and EVANS, 2003). More importantly, our data shows that those parameters are also strongly affected by effective pressure. In our experiments, the yield stress of the porous aggregates can be described by a yield surface with negative slope in plots of differential stress versus pressure. The yield surface is elliptical and is probably nonlinearly dependent on strain rate (XIAO and EVANS, 2003). In the region of negative slope, compaction rates under triaxial loads are accelerated compared to those developed under isostatic loading with the same effective mean stress. This shear-enhanced compaction indicates that deviatoric stresses provide significant inelastic contribution to the volumetric strain rate, and that the converse is also true, i.e., that shear strain rates respond to increased effective pressures. Therefore, attempts to determine $\lambda$ and $\eta$ separately without the knowledge of pressure dependence could be difficult and perhaps incorrect. Additionally, models of compaction-driven fluid flow often treat the relationship between permeability and porosity as a simple powerlaw, $k \propto \phi^n$ with a constant $n$. The data presented here suggests that neglecting the role of deformation on the connectivity and pore geometry might lead to significant errors.

Undrained shear deformation of partially molten upper mantle rocks often results in the redistribution of melt into pockets preferentially oriented with respect to the stress field. The melt segregation can be on grain scale (DAINES and KOHLSTEDT, 1997; ZIMMERMAN *et al.,* 1999), or in melt-rich bands with thicknesses and spacings larger than the grain size (HOLTZMAN *et al.,* 2003). Triaxial loads may affect the mechanical processes in fluid-saturated rocks in ways other than those discussed here. For example, JIN *et al.* (1994) suggest that dynamic wetting of grain boundaries, associated with rapid grain boundary migration, might occur during deformation. A second possibility is that isolated fluid-filled pores along grain boundaries may migration solution transfer processes (HOLNESS and GRAHAM, 1995; RUTTER, 1983). These experiments cannot test these hypotheses, particularly because solution transfer effects are in this system excluded. Clearly more work, including experiments on other mineral systems, with and without aqueous fluids or melts, and with a detailed analysis of microstructure would be desirable.

## 5. *Conclusions*

Under isostatic loading, the permeability reduction rate covaries with compaction rate and increases with increasing effective pressure. Our data also suggest that the permeability reduction rate of 10 wt% quartz samples is slightly greater than that of the 30 wt% samples. Consistent with previous studies on hot-pressed calcite aggregates, the permeability is nonlinearly related to porosity and can be fit locally with a power law $k \propto \phi^n$. However, the exponent, $n$, progressively increases as the porosity approaches a finite value, indicating a percolation porosity. When subjected to triaxial deformation, the calcite-quartz aggregates exhibit shear-enhanced compaction, but the permeability of a rock deforming under triaxial loads is not as sensitive to porosity changes as that for the same material compacting under isostatic loads: For a rock deforming under triaxial loads, $n$ seems to vary from 1.5 to 3.4. Non-isostatic deformation also reduces the percolation threshold and, in fact, enhances the permeability relative to material of the same porosity undergoing isostatic compaction.

Adding rigid second-phase particles certainly induces additional complexity to the pore structure. The addition of rigid second-phase particles also influence permeability and its evolution. Finally, our data suggest that both the bulk and shear viscosity of a saturated deformable aggregate, as commonly used in compaction theory, might be functions of both the mean and deviatoric stresses. If this conclusion is pertinent to partially molten or fluid saturated rocks in the lower crust and mantle, it may have far-reaching implications for melt extraction, deformation of fluid-laden shear zones, or compaction of granular sediments.

## *Acknowledgements*

We thank Georg Dresen, Jörg Renner, and Wenlu Zhu for thoughtful comments and constuctive reviews and gratefully acknowledge the U.S. Department of Energy for financial support under grant DOE-FG02-97ER1476. Such support does not, however, constitute an endorsement by the DOE of the views expressed in this article.

## Reference

Bernabé, Y., Brace, W. F., and Evans, B. (1982), *Permeability, porosity and pore geometry of hot-pressed calcite,* Mech. Mater. *1*, 173–183.

Bernabé, Y., Mok, U., and Evans, B. (2003), *Permeability-porosity relationships in rocks subjected to various evolution processes,* Pure Appl. Geophy. *160* (5–6), 937–960.

Bernabé, Y., Mok, U., and Evans, B. (2006), *A note on the oscillating flow method for measuring rock permeability*. Internat. J. Rock Mech. Min. Sci. *43* (2), 311–316.

Bernabé, Y., Mok, U., Evans, B., and Herrmann, F. J. (2004), *Permeability and storativity of binary mixtures of high- and low-permeability materials*, J. Geophy. Res.-Solid Earth *109* (B12).

Brace, W. F. (1980), *Permeability of crystalline and argillaceous rocks,* Int. J. Rock Mech. Min. Sci. Geomech. Abstr. *17*, 241–251.

Budiansky, B., Hutchinson, J. W., and Slutsky, S. (1982), *Void growth and collapse in viscous solids*. In (H. G. Hopkins and M. J. Sewell, Eds.), *Mechanics of Solids* (Pergamon Press, Oxford). pp. 13–45.

Connolly, J., and Podladchikov, Y. Y. (2000), *Temperature-dependent viscoelastic compaction and compartmentalization in sedimentary basins,* Tectonophysics *324*, 137–168.

Connolly, J. A. D. (1997), *Devolatilization-generated fluid pressure and deformation-propagated fluid flow during prograde regional metamorphism,* J. Geophys. Res.-Solid Earth *102* (B8), 18149–18173.

Connolly, J. A. D. and Podladchikov, Y. Y. (2004), *Fluid flow in compressive tectonic settings: Implications for midcrustal seismic reflectors and downward fluid migration,* J. Geophy. Res.-Solid Earth *109* (B4).

Daines, M. J. and Kohlstedt, D. L. (1993), *A laboratory study of melt migration,* Phil. Trans. R. Soc. London A, *342*, 43–52.

Daines, M. J. and Kohlstedt, D. L. (1997), *Influence of deformation on melt topology in peridotites,* J. Geophys. Res., B, Solid Earth and Planets *102* (5), 10,257–10,271.

De Bresser, J. H. P. and Spiers, C. J. (1993), *Slip systems in calcite single crystals deformed at 300–800 degrees C.* J. Geophys. Res. B, Solid Earth and Planets *98* (4), 6397–6409

Etheridge, M. A., Wall, V. J., Cox, S. F., and Vernon, R. H. (1984), *High fluid pressures during regional metamorphism and deformation: Implications for mass transport and deformation mechanisms,* J. Geophys. Res. *89*, 4344–4358.

Evans, B., Bernabé, Y., and Zhu, W. (1999), *Evolution of pore structure and permeability of rocks during laboratory experiments.* In (B. Jamtveit and P. Meakin, Eds), *Growth, Dissolution, and Pattern Formation in Geosystems* (Kluwer Academic Publishers), pp. 327–344.

Faulkner, D. R. and Rutter, E. H. (2000), *Comparisons of water and argon permeability in natural clay-bearing fault gouge under high pressure at 20 degrees* C, J. Geophys. Res.-Solid Earth *105* (B7), 16415–16426.

FAULKNER, D. R., and RUTTER, E. H. (2003), *The effect of temperature, the nature of the pore fluid, and subyield differential stress on the permeability of phyllosilicate-rich fault gouge,* J. Geophys. Res.-Solid Earth *108* (B5).

FISCHER, G. J. (1992), *The determination of permeability and storage capacity: Pore pressure oscillation method.* In (B. EVANS and T.-f. WONG, Eds.), *Fault Mechanics and Transport Properties of Rocks; a Festschrift in Honor of W. F. Brace*, pp. 187–212.

FISCHER, G. J. and PATERSON, M. S. (1992), *Measurement of Permeability and Storage Capacity in Rocks During Deformation at High Temperature and Pressure.* In (B. EVANS and T.-f. WONG, Eds.), *Fault Mechanics and Transport Properties of Rocks; a Festschrift in Honor of W. F. Brace*, pp. 213–252.

FISHER, D. M. and BRANTLEY, S. L. (1992), *Models of quartz overgrowth and vein formation; deformation and episodic fluid flow in an ancient subduction zone,* J. Geophys. Res. B, Solid Earth and Planets *97* (13), 20,043–20,061.

GUÉGUEN, Y. and DIENES, J. (1989), *Transport Properties of Rocks From Statistics and Percolation.,* Math. Geol. *21,* 1–13.

HEILAND, J. and RAAB, S. (2001), *Experimental investigation of the influence of differential stress on permeability of a Lower Permian (Rotliegend) sandstone deformed in the brittle deformation field,* Phys. Chem. Earth, Part A-Solid Earth and Geodesy *26* (1–2), 33–38.

HELLE, A. S., EASTERLING, K. E., and ASHBY, M. F. (1985), *Hot-isostatic pressing diagrams: New developments*, Acta Metall. *33,* 2163–2174.

HICKMAN, S. H. and EVANS, B. (1987), *Influence of geometry upon crack healing rate in calcite,* Phys. Chem. Minerals *15* (1), 91–102.

HOLCOMB, D. J. and OLSSON, W. A. (2003), *Compaction localization and fluid flow,* J. Geophys. Res.-Solid Earth *108* (B6).

HOLNESS, M. B. and GRAHAM, C. M. (1995), *P-T-X Effects on equilibrium carbolate-$H_2$o-$Co_2$nacl dihedral angles – constraints on carbonate permeability and the role of deformation during fluid infiltration,* Contributions to Mineralogy and Petrology, *119* (2–3), 301–313.

HOLTZMAN, B. K., GROEBNER, N. J., ZIMMERMAN, M. E., GINSBERG, S. B., and KOHLSTEDT, D. L. (2003), *Stress-driven melt segregation in partially molten rocks,* Geochem. Geophy. Geosystems 4.

JIN, Z.-M., GREEN, H. W., and ZHOU, Y. (1994), *Melt topology in partially molten mantle peridotite during ductile deformation,* Nature *372,* 164–167.

MAIN, I. G., KWON, O., NGWENYA, B. T., and ELPHICK, S. C. (2000), *Fault sealing during deformation-band growth in porous sandstone,* Geology *28* (12), 1131–1134.

MCKENZIE, D. (1984), *The generation and compaction of partially molten rock,* J. Petrol. *25,* 713–765.

MCKENZIE, D. and HOLNESS, M. (2000), *Local deformation in compacting flows: development of pressure shadows,* Earth Planet. Sci. Lett. *180* (1–2), 169–184.

MCKENZIE, D. P. (1987), *The compaction of Igneous and sedimentary-rocks,* J. Geolog. Soc. *144,* 299–307.

NGWENYA, B. T., KWON, O., ELPHICK, S. C., and MAIN, I. G. (2003), *Permeability evolution during progressive development of deformation bands in porous sandstones,* J. Geophys. Res.-Solid Earth *108* (B7).

NUR, A. and WALDER, J. (1992), *Hydraulic pulses in the Earth's crust,* In (B. Evans and T.-f. Wong, Eds.), *Fault Mechanics and Transport Properties of Rocks.* (Academic Press, San Diego, CA, 1992), pp. 461–473.

OJALA, I. O., NGWENYA, B. T., MAIN, I. G., and ELPHICK, S. C. (2003), *Correlation of microseismic and chemical properties of brittle deformation in Locharbriggs sandstone,* J. Geophys. Res.-Solid Earth *108* (B5).

PATERSON, M. S. (1995), *A theory for granular flow accommodated by material transfer via an intergranular fluid,* Tectonophysics *245*, 133–151.

PEACH, C. J. and SPIERS, C. J. (1996), *Influence of crystal plastic deformationon dilatancy and permeability development in synthetic salt rock,* Tectonophysics *256,* 101–128.

POPP, T., KERN, H., and SCHULZE, O. (2001), *Evolution of dilatancy and permeability in rock salt during hydrostatic compaction and triaxial deformation,* J. Geophys. Res.-Solid Earth *106* (B3), 4061–4078.

RENNER, J., VISKUPIC, K., HIRTH, G., and EVANS, B. (2003), *Melt extraction from partially molten peridotites*, Geochem. Geophys. Geosystems 4.

RENNER, R., EVANS, B., and Siddiqi, G. (2002), *Dislocation creep of calcite,* J. Geophys. Res.-Solid Earth *107* (B12).

RUTTER, E. (1997), *The influence of deformation on the extraction f crustal melts: a consideration of the role of melt-assisted granular flow,* In (M. B. HOLNESS, Ed.), *Deformation-enhanced Fluid Transport in the Earth's Crust and Mantle.* (Chapman and Hall, London 1997), pp. 82–110.

RUTTER, E. H. (1974), *The influence of temperature, strain rate and interstitial water in the experimental deformation of calcite rocks,* Tectonophysics *22*, 311–334.

RUTTER, E. H. (1983), *Pressure solution in nature, theory and experiment,* J. Geolog. Soc. *140* (SEP), 725–740.

RYBACKI, E., PATERSON, M. S., WIRTH, R., and DRESEN, G. (2003), *Rheology of calcite-quartz aggregates deformed to large strain in torsion,* J. Geophys. Res.-Solid Earth *108* (B2).

SCOTT, D. R. (1988), *The competition between percolation and circulation in a deformable porous-medium,* J. Geophys. Res.-Solid Earth and Planets *93* (B6), 6451–6462.

SCOTT, D. R., and STEVENSON, D. J. (1984), *Magma Solitons.* Geophys. Res. Lett. *11* (11), 1161–1164.

SPIEGELMAN, M. (1993), *Flow in deformable porous-media. 1. Simple analysis*, J. Fluid Mech. *247*, 17–38.

STEVENSON, D. J. (1989), *Spontaneous Small-scale Melt Segregation in Partial Melts Undergoing Deformation*, Geophys. Res. Lett. *16* (9), 1067–1070.

UEHARA, S. and SHIMAMOTO, T. (2004), *Gas permeability evolution of cataclasite and fault gouge in triaxial compression and implications for changes in fault-zone permeability structure through the earthquake cycle*, Tectonophysics *378* (3–4), 183–195.

VAJDOVA, V., BAUD, P., and WONG, T. F. (2004), *Permeability evolution during localized deformation in Bentheim sandstone,* J. Geophys. Res.-Solid Earth *109* (B10).

VAN BARGEN, N. and WAFF, H. S. (1986), *Permeabilities, interfacial areas and curvature of partially molten systems: Results of numerical computations of equilibrium microstructures,* J. Geophys. Res. *91*, 9261–9276.

WALDER, J. and NUR, A. (1984), *Porosity reduction and crustal pore pressure development,* J. Geophys. Res. *89*, 11539–11548.

WALSH, J. B. (1965), *The effect of cracks on the uniaxial elastic compression of rocks,* J. Geophys. Res. *70* (2), 399–411.

WARK, D. A. and WATSON, E. B. (1998), *Grain-scale permeabilities of texturally equilibrated, monomineralic rocks,* Earth Planet. Sci. L. *164*, 591–605.

WATSON, E. B. and BRENAN, J. M. (1987), *Fluids in the lithosphere, 1. Experimentally-determined wetting characteristics of $CO_2$-$H_2O$ fluids and their implications for fluid transport, host-rock physical properties, and fluid inclusion formation,* Earth Planet. Sci. Lett. *85*, 497–515.

WILKINSON, D. S. and ASHBY, M. F. (1975), *Pressure sintering by power-Law Creep* Acta Metallurgica *23* (11), 1277–1285.

WONG, T.-f. (1990), *Mechanical compaction and the brittle-ductile transition in porous sandstones,* In (R. J. Knipe and E. H. Rutter Eds.), *Deformation Mechanisms, Rheology and Tectonics*, Geolog. Soc. Special Publ. *54* (Geologi. Soc., London 1990) pp. 111–122.

XIAO, X. H. and EVANS, B. (2003), *Shear-enhanced compaction during non-linear viscous creep of porous calcite-quartz aggregates.* Earth Planet. Sci. Letts. *216* (4), 725–740.

ZHANG, S. Q., COX, S. F., and PATERSON, M. S. (1994a), *The influence of room-temperature deformation on porosity and permeability in calcite aggregates.* J. Geophys. Res.-Solid Earth *99* (B8), 15761–15775.

ZHANG, S. Q., FITZ GERALD, J. D., and COX, S. F. (2000), *Reaction-enhanced permeability during decarbonation of calcite plus quartz wollastonite plus carbon dioxide,* Geology *28* (10), 911–914.

ZHANG, S. Q., PATERSON, M. S., and COX, S. F. (1994b), *Porosity and permeability evolution during hot isostatic pressing of calcite aggregates,* J. Geophys. Res.-Solid Earth *99* (B8), 15741–15760.

ZHANG, S. Q., PATERSON, M. S., and COX, S. F. (2001), *Microcrack growth and healing in deformed calcite aggregates,* Tectonophysics *335* (1–2), 17–36.

ZHU, W. L., EVANS, B., and BERNABÉ, Y. (1999), *Densification and permeability reduction in hot-pressed calcite: A kinetic model,* J. Geophys. Res.-Solid Earth *104* (B11), 25501–25511.

ZHU, W. L., MONTESI, L. G. J., and WONG, T. F. (1997), *Shear-enhanced compaction and permeability reduction: Triaxial extension tests on porous sandstone,* Mech. Mater. *25* (3), 199–214.
ZHU, W. L. and WONG, T. F. (1997), *The transition from brittle faulting to cataclastic flow: Permeability evolution,* J. Geophys. Res.-Solid Earth *102* (B2), 3027–3041.
ZIMMERMAN, M. E., ZHANG, S. Q., KOHLSTEDT, D. L., and KARATO, S. (1999), *Melt distribution in mantle rocks deformed in shear*, Geophys. Res. Lett. *26* (10), 1505–1508.

(Received November 8, 2005, revised March 15, 2006, accepted March 17, 2006)

Pure appl. geophys. 163 (2006) 2103–2123
0033–4553/06/102103–21
DOI 10.1007/s00024-006-0121-3



Pure and Applied Geophysics

# Experimental Investigation into the Scale Dependence of Fluid Transport in Heterogeneous Rocks

INSUN SONG and JÖRG RENNER

*Abstract*—We investigated the dependence of hydraulic properties on the spatial scale of intrinsic and artificial heterogeneity, applying harmonic pore pressure testing to two varieties of Fontainebleau sandstone at various periods and effective pressures. Blocks with porosities of about 5 and 8% were chosen exhibiting a permeability of about $2{\cdot}10^{-19}$ and $1{\cdot}10^{-13}$ $m^2$, respectively. The permeability of the less permeable variety strongly depends on sample size. Artificial heterogeneous rock samples were prepared by stacking pieces of the two sandstone varieties perpendicular and parallel to the main flow direction. The perturbation of the fluid flow owing to the interfaces between pieces of the same variety is controlled by the orientation of and subordinately by the effective stress on the interfaces. Constraints on conduit geometry derived from the effect of interfaces indicate that interconnectivity is more important than pore radius at the lower porosity. The effective permeability of alternating stacks of the two varieties differs tremendously for the two interface orientations; arithmetic and harmonic averages coincide with the effective properties parallel and perpendicular to the main flow, respectively. When the oscillation period is varied two regimes are observed, one with constant permeability at long periods and a second with decreasing permeability for decreasing period at short periods. Order of magnitude considerations of penetration depth suggest that this period dependence may be related to heterogeneity.

**Key words:** Fluid transport, dynamic permeability, scale-dependent transport properties, effective medium, percolation threshold, Fontainebleau sandstone.

## 1. *Introduction*

The investigation of fluid transport in geological material is of fundamental importance for many practical problems including petroleum exploration, fresh-water supply, environmental remediation, and hazardous waste disposal. A standard approach to characterize the efficiency of fluid flow through rocks relies on Darcy's law

$$\vec{q} = -\frac{\mathbf{k}}{\mu}\vec{\nabla}p_f, \tag{1}$$

Institute for Geology, Mineralogy, and Geophysics, Ruhr-University Bochum, D-44780, Bochum, Germany. E-mail: insun.song@rub.de; renner@geophysik.rub.de which states that the amount of fluid flowing through a unit cross section per unit time, $\vec{q}$, is proportional to the gradient of fluid pressure, $\vec{\nabla} p_f$. The proportionality constant is separated into the bulk rock property, the tensor of hydraulic permeability $\mathbf{k}$ [m$^2$], and the fluid viscosity $\mu$ [Pa s]. From a theoretical perspective, the Navier-Stokes equations, in which permeability is not an explicit parameter, describe the relative motion between fluid and solid components in a porous medium. The solid frame imposes boundary conditions on the fluid transport that are difficult to quantify explicitly since the geometry of the pore surface and its displacement have to be exactly described. Yet, Darcy's law can be derived from the Navier-Stokes equations relying on statistical descriptions of the boundary conditions (e.g., HILFER, 1996).

The Navier-Stokes equations for a porous medium contain viscous and inertial forces (e.g., PRIDE, 2005). With increasing flow velocity, inertia effects become significant, i.e., the local flow occurs at larger than critical Reynold's numbers, causing deviations from the linear law (1) or a frequency dependence of permeability for harmonic excitation. Besides inertia effects, a change in dominating fluid path with flow velocity may explain deviations from the linear relation (ANDRADE *et al.*, 1997). The consideration of harmonic hydrodynamic problems led to the concept of dynamic permeability (JOHNSON *et al.*, 1987). Defining a complex-valued dynamic permeability constitutes a compact account of the frequency-controlled switch from viscosity-controlled diffusion to inertia-controlled wave propagation (e.g., CHARLAIX *et al.*, 1988; JOHNSON *et al.*, 1987; SHENG and ZHOU, 1988). The transition period, $T_c$, between dynamic and static permeability marks the cross-over of two length scales, the characteristic pore radius and the viscous skin depth, $\delta = \sqrt{\mu T/(\pi\rho)}$, the characteristic size of the fluid layer in which transport is controlled by viscous forces. Here, $\rho$ denotes fluid density.

Darcy's law (1) is most easily tested by steady-state flow-through experiments. The advent of transient flow methods addressed experimental problems associated with tight rocks (e.g., BRACE *et al.*, 1968). Oscillatory testing was initially introduced in physical chemistry (TURNER, 1958, 1959) and then applied to rock samples (FISCHER, 1992; KRANZ *et al.*, 1990; STEWART *et al.*, 1961). Frequency dependence of derived permeability values was previously reported from experiments using the oscillatory method (BERNABÉ *et al.*, 2004; FISCHER and PATERSON, 1992). The frequency dependence may in principle be related to the described transition from laminar to turbulent flow, to nonlinear pressure sensitivity of sample properties, to sample heterogeneity, and to differences between the assumed and actually realized boundary conditions.

The primary objective of our study is to explore to what extent sample heterogeneity actually causes frequency dependence of permeability and whether a relation can be found between the spatial scale of heterogeneity and the time scale of the oscillation. Toward this end we conducted a series of oscillating pore pressure tests to determine the permeability of two varieties of Fontainebleau sandstone at various periods and pressure conditions. Macroscopically heterogeneous samples were fabricated with artificial interfaces parallel and perpendicular to the main flow or composed of stacks of the two different varieties. The two alternative causes for a period dependence of permeability, pressure sensitivity and poor realization of boundary conditions, were addressed by choosing samples with pressure insensitive characteristics and optimizing the sample assembly, respectively.

## 2. *Experimental Approach*

### 2.1. *Sample Selection and Preparation*

Fontainebleau sandstone, a clay-free, ultra-pure (≈100% quartz), well sorted sandstone of the Oligocene age found in the Ile de France region around Paris, has been extensively used for investigating the relation between transport properties and pore structure (BOURBIE and ZINSZNER, 1985; DAVID and DAROT, 1989; DOYEN, 1988; FREDRICH *et al.*, 1993). Bulk porosity and the characteristics of pore structure widely vary without significant grain size modification. Eight blocks of Fontainebleau sandstone from different locations were available for our laboratory tests, ranging in porosity from about 3 to 15% with moderate variation in a given block (Fig. 1a). A difference between total and open porosity could not be resolved within experimental uncertainty (except for block FS1 with porosity of ~12%). Note, open porosity was determined from imbibition of samples that were entirely submersed in water upon evacuation. Thus, this open porosity is not necessarily identical to the amount of pore space contributing to the directional interconnectivity from one end face to the other of the cylindrical samples relevant for the flow induced by the pore pressure oscillations. Measured ultrasonic P-wave velocities in water-saturated samples cover the entire range between the lower and the upper Hashin-Shtrikman bounds (HASHIN and SHTRIKMAN, 1963). A pilot study revealed that permeability values determined by steady-state flow at effective pressures less than ~10 MPa significantly decrease with decreasing porosity. The drop is most prominent for porosities less than ~8% (Fig. 1b) in agreement with previous observations (BOURBIE and ZINSZNER, 1985).

For our investigation, two blocks, FS7 and FS8, were chosen with average porosities of 8.2 and 5.3%, i.e., close to and below the critical value of 8%, respectively, and the highest ultrasonic P-wave velocities (Fig. 1a). Based on variational principles for effective elastic properties, it may be concluded that the pore structure of these two blocks contains only a subordinate amount of compliant crack-like conduits. Microcracks are also not detected with the scanning electron microscope (Figs. 1c, d). In fact, grain boundaries are so tight that they can hardly be discerned. Quantitative image analysis of optical micrographs of thin sections and SEM images of polished surfaces yield statistical two-dimensional averages of grain and pore size and coordination number (Table 1).

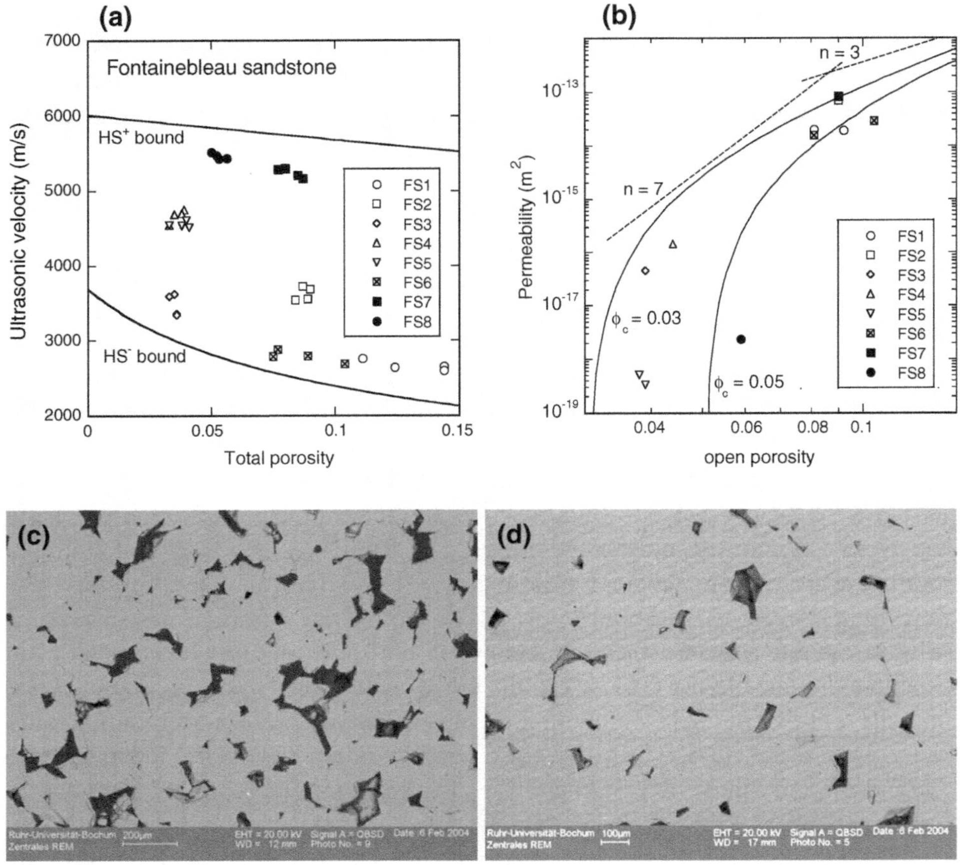


Figure 1
Properties of water-saturated, cylindrical samples with 30 mm diameter and 60 mm length from eight blocks of Fontainebleau sandstone: a) Ultrasonic P-wave velocity as varying with total porosity. Lines indicate the lower ($HS^-$) and upper ($HS^+$) Hashin-Shtrikman bounds for isotropic two-phase mixtures. b) Permeability as varying with total porosity determined by steady-state flow at an effective pressure below ~ 10 MPa. Solid curves indicate modeling of the data using eq. (5) with $n = 3$ and $\phi_c = 0.03$ and 0.05. Dashed lines indicate the bilinear approximation of the results of Bourbie and Zinszner (1985). c) and d) Scanning electron microscope images of varieties FS7 and FS8, respectively. Individual quartz grains (quartz: light grey; pores: dark grey) are difficult to discern suggesting tight grain contacts.

Cylindrical samples with a diameter of 30 mm were drilled from the blocks in a consistent orientation. Sample ends were sawed and ground to be square. Sample length amounted to 10, 20, 30, 60, and 110 mm. We modeled heterogeneous rock samples by stacking pieces of the two varieties to gain samples of 60 mm length in two ways, perpendicular and parallel to the main flow direction, the cylinder axis (Fig. 2). Samples with interfaces parallel to the flow direction were drilled from stacks of ground rock plates with the appropriate thickness. We used the same

Table 1

*Physical properties of the two tested Fontainebleau sandstone varieties as determined for samples with a length of 60 mm and a diameter of 30 mm. Statistical parameters of the microstructure were determined from two-dimensional sections and are reported without stereological corrections*

| Block | Total porosity $\phi$ (%) | Permeability $k_0$ (m$^2$) | Formation factor $F^1$ (–) | Critical radius$^2$ (μm) | Transition period $T_c$ (s) | Critical skin depth$^3$(mm) | Critical penetration depth$^4$ (m) | Average grain size $\bar{d}^5$ (μm) | Average pore size$^6$ (μm) | Coordination number$^7$ (–) |
|---|---|---|---|---|---|---|---|---|---|---|
| FS7 | 8.2 ± 0.5 | $1.3 \cdot 10^{-13}$ | 88 ± 16 | 17 | 0.1 | 0.18 | 0.6 | 130 ± 52 | 25 – 30 | 4.5 ± 1.3 |
| FS8 | 5.3 ± 0.3 | $2.0 \cdot 10^{-19}$ | 156 ± 23 | 0.03 | 500 | 13 | 0.06 | 205 ± 66 | 25 – 30 | 5.3 ± 1.2 |

[1] $F = \frac{\sigma_f}{\sigma}$, determined from separate measurements of the electrical conductivity $\sigma$ and $\sigma_f$ of saturated samples and pore fluid, respectively.

[2] $r_c \simeq \sqrt{25 k_0 F}$ (e.g., BERNABÉ and BRUDERER, 1998).

[3] $\delta_c = \sqrt{\frac{\eta T_c}{\pi \rho}}$ (e.g., ZHOU and SHENG, 1989).

[4] according to eq. (6).

[5] average diameter of circle with equivalent area as grain.

[6] most frequent diameter of circle with equivalent area as pore, analysis was performed on images with magnification similar to the ones presented in Figures 1c and d., i.e., the lower detection limit for pores is on the order of a few microns.

[7] number of grains in contact with a grain.

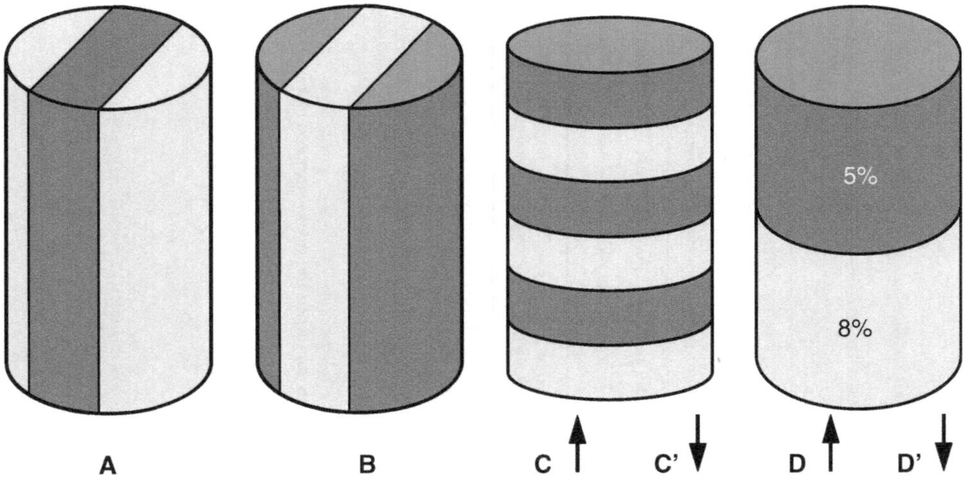


Figure 2
Labeling for mixed stacks. The lighter and darker grey colors represent the variety FS7 and FS8 with ~8 and 5% porosity, respectively. Arrows indicate flow direction.

grinding procedure for all surfaces finishing with 800 mesh. Stacks were composed of pieces of one variety but also of the two varieties alternating.

## *2.2. Experimental Procedure*

Our test arrangement consisted of a specimen placed in a confining pressure vessel and connected to a downstream and an upstream reservoir equipped with servo-hydraulically controlled intensifiers (see SONG and RENNER, 2006). Porous metal discs (SIKAR S) of 3 mm thickness with a porosity of 31%, a permeability of $> 10^{-12}$ m$^2$, and a pore diameter in the range of 7 to 14 μm at the sample ends served a dual purpose. Firstly, pore fluid is evenly distributed over the sample end faces, and secondly, the confining pressure is transmitted to the sample in axial direction, too. The water-saturated rock sample with the porous discs and end plugs was held together by a rubber jacket that also separated confining and pore fluid. Constant confining pressure of up to 200 MPa was applied. Using the pressure intensifier, the pore pressure was raised to a prescribed level in the entire pore pressure system. We define effective pressure as the difference between confining and pore-fluid pressure, $p_{\mathrm{eff}} = p_c - p_f$. After equilibration of pore pressure in the specimen, the two reservoirs were separated and a sinusoidal oscillation of the upstream pressure was generated that is transmitted through the sample and causes retarded and attenuated sinusoidal oscillations of the downstream pressure. Water viscosity was calculated as a function of pressure (http://www.ruhr-uni-bochum.de/thermo/Software/Seiten/Fluidcal-eng.htm) but temperature variations were not taken into account. No precautions were taken to maintain the apparatus at a constant temperature but variations remained less than 5°C during a night-day cycle and thus may have a moderate effect on pore fluid density and viscosity only for the tests with the longest applied oscillation periods of 33 minutes.

Tests on a single sample were performed at several different pore-fluid pressure levels up to 180 MPa and at periods varying up to three orders between 0.04 and 2000s. For a specific sample, variations in oscillation period were limited toward short periods by the detection limit of the downstream pressure gauge and toward long periods by the vanishing difference between the pressures in the two reservoirs. We present data as a function of effective pressure though confining pressure was varied, too, because deviations from a simple effective pressure law for permeability are subordinate compared to the effects of effective pressure and period variations for the investigated sandstone varieties with limited pore compressibility.

### *2.3. Data Processing*

The derivation and characteristics of the analytical solution of the one-dimensional pressure diffusion equation with the boundary conditions of oscillating upstream pressure and a downstream reservoir of finite size have been presented before (BERNABÉ *et al.*, 2006; FISCHER, 1992; KRANZ *et al.*, 1990). The solution is valid for a homogeneous isotropic porous medium with a linear constitutive equation for the storage capacity arising from fluid and frame compressibility. It predicts an amplitude ratio or attenuation, $\alpha$, and a phase lag, $\delta$, between upstream and downstream pressure of

$$\alpha = \left| \left( \cosh\left[ (1+i)\sqrt{\frac{\xi}{\eta}} \right] + \frac{1+i}{\sqrt{\xi\eta}} \sinh\left[ (1+i)\sqrt{\frac{\xi}{\eta}} \right] \right)^{-1} \right| \tag{2}$$

and

$$\delta = \arg\left[ \left( \cosh\left[ (1+i)\sqrt{\frac{\xi}{\eta}} \right] + \frac{1+i}{\sqrt{\xi\eta}} \sinh\left[ (1+i)\sqrt{\frac{\xi}{\eta}} \right] \right)^{-1} \right], \tag{3}$$

where $\eta$ and $\xi$ denote dimensionless permeability and storage capacity, respectively. It is important to note, that the used analytical solution remains valid when a complex-valued dynamic permeability is introduced in the governing equations to account for fluid inertia. Furthermore, the notion that Darcy's law is valid on a scale much smaller than sample size is inherent to the solution.

The two parameters, $\eta$ and $\xi$, were determined using a grid search algorithm minimizing the error between the measured and calculated attenuation and phase shift values. For an applied oscillation period $T$, sample permeability can then be determined as

$$k = \frac{\pi \mu L B_d}{TA} \eta, \tag{4}$$

where $\mu$, $B_d$, $L$, and $A$ denote the dynamic viscosity of the fluid, the storage capacity of the downstream reservoir, the length and the cross-sectional area of the sample, respectively. The specific storage capacity calculates according to $\beta_s = \xi B_d/(AL)$ and the hydraulic diffusivity $D = \pi L^2 \eta/(T\xi)$. However, in this presentation we exclude our results for storage capacity because unresolved problems currently remain regarding the accuracy of storage capacity determined from oscillatory tests (BERNABÉ *et al.*, 2006).

## 3. Results

### 3.1. Intact Samples

Four intact samples, three 60 mm and one 110 mm long, from the block with ~8% porosity (FS7) were tested at the same range of effective pressures and oscillation periods yielding a permeability of $\sim 10^{-13}$ m$^2$. Results of the oscillatory tests are in good agreement with steady-state measurements in which the variation in effective pressure was achieved by changing confining pressure alone and pore pressure was maintained at 5 MPa and at ambient conditions at the two sample ends. Permeability is insensitive to variations in effective pressure (Fig. 3a), confirming the suggestion based on ultrasonic measurements and microscopic analysis that the sandstone samples are devoid of compliant crack-like conduits. We also did not find a significant effect of oscillation amplitude between 1 and 8 MPa. A transition from period-sensitive to period-insensitive behavior is documented for the 110 mm long sample (Fig. 3b). The plateau value reached

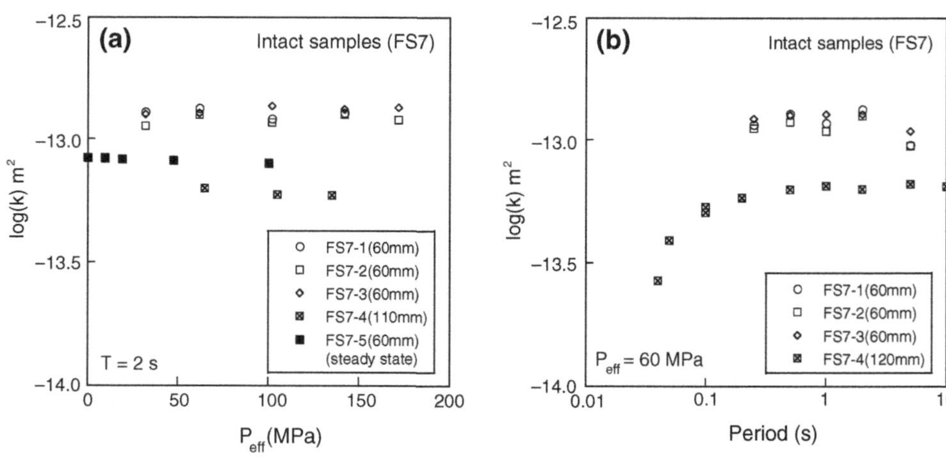


Figure 3
Permeability of intact samples of variety FS7 with a porosity of ~8% as a function of a) effective pressure and b) period.

above periods of ~0.1 s is slightly lower than for the shorter samples. Note, that oscillatory measurements on samples shorter than 60 mm were not feasible, because phase shift and attenuation between up- and downstream pressure could not be resolved even at the shortest periods.

Permeability of a 60 mm long sample of the sandstone variety with ~5% porosity (FS8) increases with a decreasing rate as the period is raised. A plateau of constant values of $\sim 2 \cdot 10^{-19}$ $m^2$ is approached at periods longer than ~500 s (Fig. 4). Reducing the length to 30 mm hardly affects permeability but allows us to confirm the decrease in permeability with decreasing period. For a sample length below 30 mm, the characteristics of permeability dramatically change. First, absolute values are 100 to 1000 times higher than those of longer samples but still about three orders of magnitude smaller than for the variety with ~8% porosity. Second, a weak increase in permeability with decreasing period gives way to a reversed relation at periods below ~1 s for two samples. The moderate dependence on effective pressure

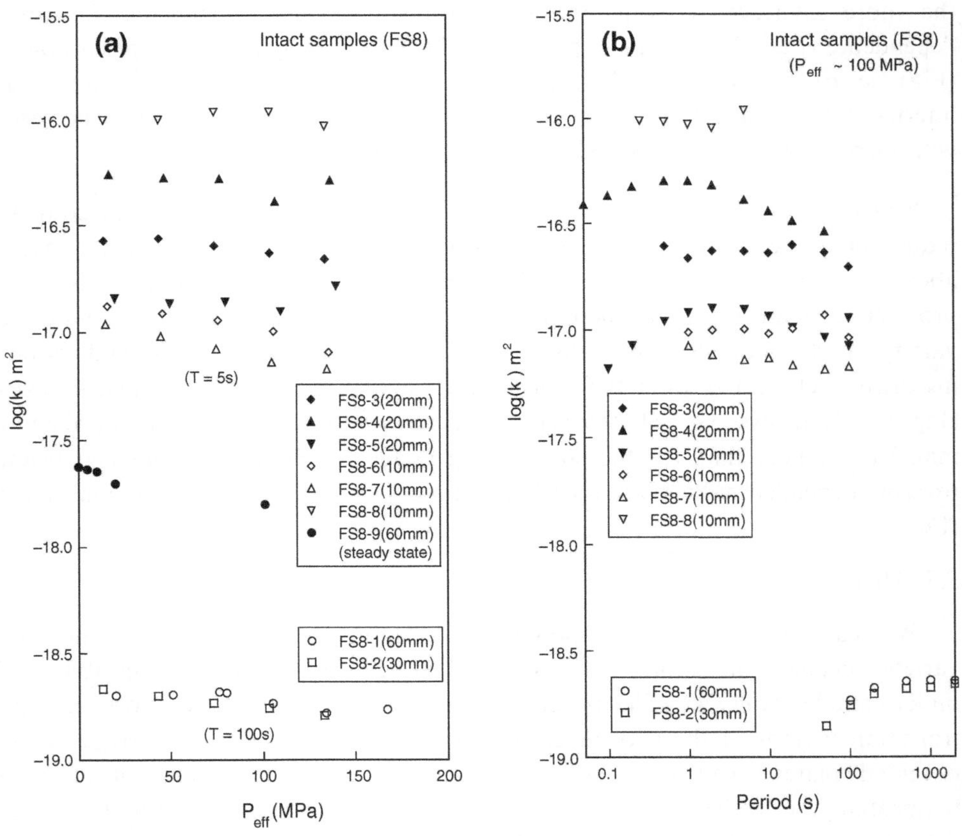


Figure 4

Permeability of intact samples of variety FS8 with a porosity of ~5% as a function of a) effective pressure and b) period. The length of the tested samples is indicated in the legend.

is comparable to that for longer samples. The results of steady-state flow-through measurements on a sample with 60 mm length fall between the permeability deduced from oscillatory tests on samples with lengths above and below 30 mm. Since we did not test the same sample with both methods, it remains speculative whether the steady-state method yields an averaged permeability compared to the oscillatory method. Porosity of the sample used in the steady-state experiment is at the upper end of values for the block and the difference to the oscillatory measurement could solely be due to variations in porosity (Fig. 1b).

### *3.2. Stacks of a Single Variety*

Interfaces perpendicular and parallel to the main flow direction affect permeability differently.

*Variety with ~8% porosity (FS7):* The average permeability decreases with increasing number of interfaces perpendicular to the main flow direction (Fig. 5a); the more interfaces present, the more pronounced is the effective pressure-dependence (Fig. 5b). Interfaces parallel to the main flow hardly affect permeability (Figs. 5c, d). In particular, there is no systematic dependence on the number of interfaces. Independent of interface orientation, permeability continuously increases with increasing oscillation period in contrast to intact samples.

*Variety with ~5% porosity (FS8):* Two perpendicular interfaces substantially reduce the permeability compared to the constituting 20 mm long pieces, but it remains above that of the 60 mm long sample (Fig. 6). The pressure dependence is more pronounced than for any of the intact samples (Fig. 6b) or samples containing the variety with 8% porosity. The transition period between period-sensitive and period-insensitive behavior is lower than for long but higher than for short intact samples (Fig. 6a). With two parallel interfaces, the permeability is more than two orders of magnitude higher than for the 60 mm long intact sample and exhibits significant pressure dependence; the period dependence becomes complex with sigmoidal shape (Fig. 6).

### *3.3. Mixed Stacks*

Permeability of samples composed of alternating stacks of the two sandstone varieties depends on oscillation period qualitatively similar to the permeability of intact samples (Fig. 7a). The plateau values are reached at periods between the transition periods of the two intact end-members (Fig. 7b). In comparison to the observed plateau values of mixed samples, we calculated values of effective permeability. Effective transport properties of a heterogeneous material may be determined from volume-fraction weighted averages of properties of the constituting phases (e.g., MAVKO *et al.*, 1998; RENARD and DE MARSILY, 1997; TORQUATO, 2002). The specific relation between the effective and the constituent's properties, the mixing

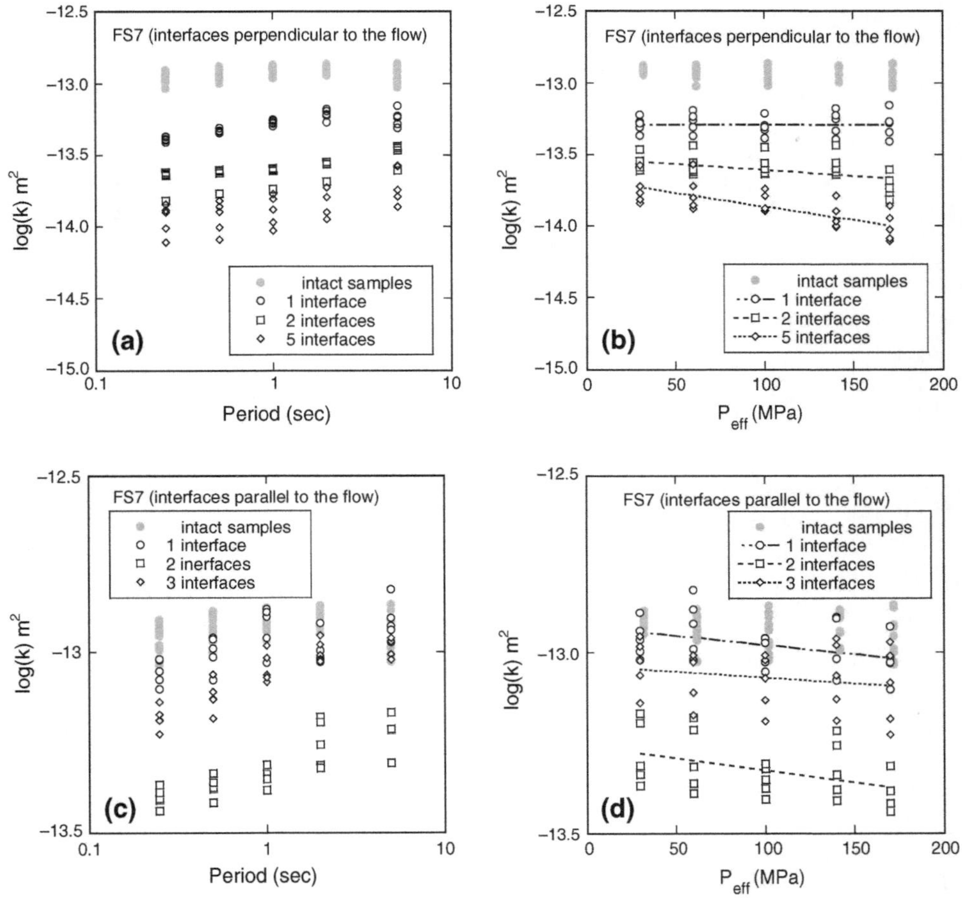


Figure 5
Effect of interfaces perpendicular (a and b) and parallel (c and d) to the cylinder axis of samples composed of the variety with ~8% porosity (FS7) on the period and pressure dependence of permeability. The variation of permeability at constant effective pressure results from variations in oscillation period and *vice versa*. Grey symbols reproduce the results for intact samples for comparison.

rule, strongly depends on the structural characteristics as exemplified by the variation in ultrasonic velocity for given phase proportion, i.e., porosity (Fig. 1a). Owing to the simplicity of the geometrical mixture of the stacks, we considered two relations, the arithmetic average $\bar{k}_a = \sum_{I=1}^{n} x_i k_i$ and the harmonic average $1/\bar{k}_h = \sum_{i=1}^{n} x_i/k_i$ of permeability $k_i$ of the i-th constituent with volume fraction $x_i$. The arithmetic average matches the experimental results for main flow parallel to the layering, while the harmonic average predicts the effective properties for main flow perpendicular to the layering (Fig. 8). For the calculations, we assumed that sub-samples of the high

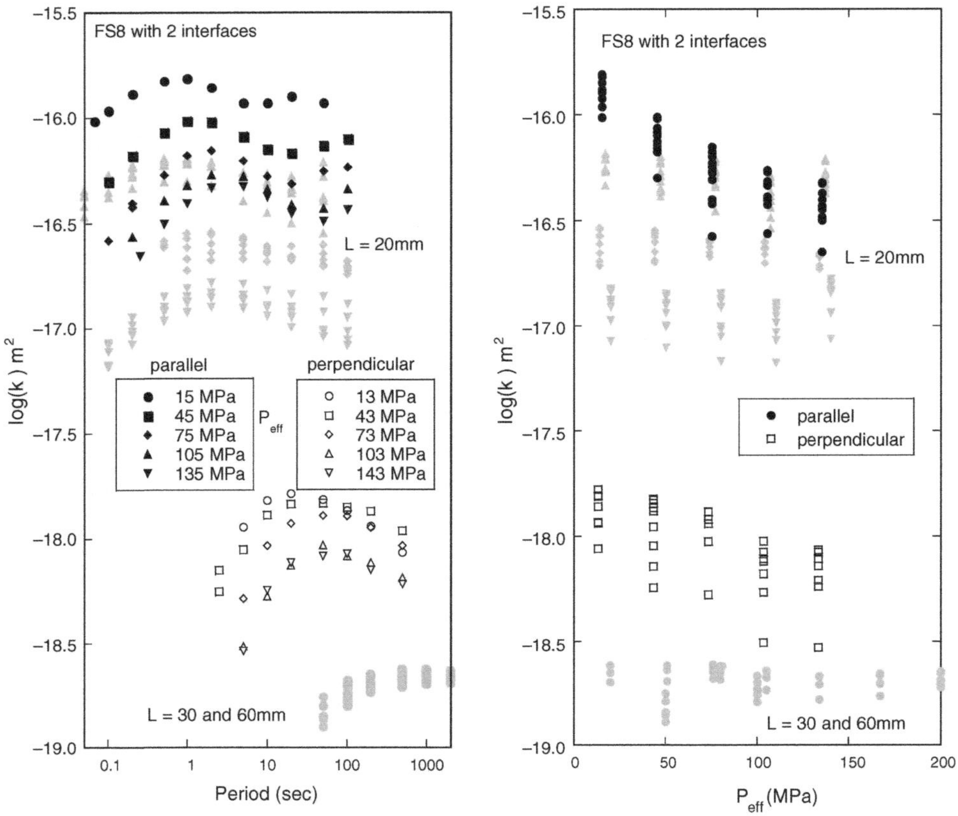


Figure 6
Effect of interfaces perpendicular and parallel to the cylinder axis of samples composed of the variety with ~5% porosity (FS8) on (a) the period and (b) the pressure dependence of permeability. The variation of permeability at constant effective pressure results from variations in oscillation period. Grey symbols reproduce the results for intact samples for comparison.

porosity variety (FS7) have the same permeability independent of the involved length scales. The properties of low porosity samples (FS8), however, vary with length. In this case, we relied on the measurements of the permeability of each cylindrical sub-sample. The slender pieces resulting from flow-parallel cuts were assumed to have the same properties as intact samples with the same length.

## *4. Discussion*

We observed distinct effects of the length for intact samples and of the presence of interfaces. Furthermore, permeability of intact and stacked samples is affected by oscillation period. In the following discussion, we focus on the

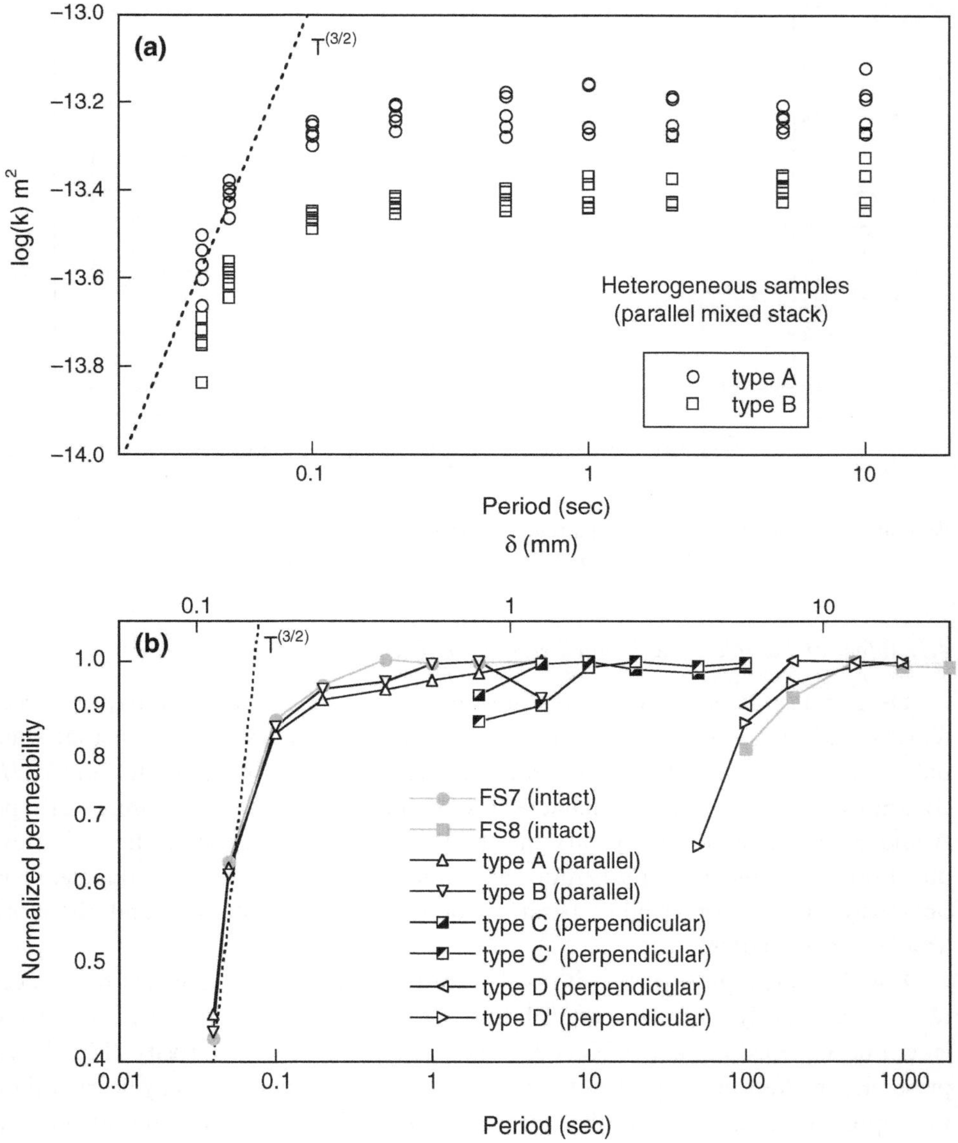


Figure 7
a) Period dependence of the permeability of samples with flow-parallel layering (types A and B; see Figure 2). b) Compilation of period dependence of relative permeability for mixed samples. Permeability values were normalized to the plateau values reached at long periods. The dashed lines indicate the relation $k \propto T^{3/2}$ suggested by scaling theories for dynamic permeability.

potential relation between the effect of sample length and percolation theory, the period dependence and heterogeneity, and on constraints on pore-network geometry that can be derived from the properties of stacked samples.

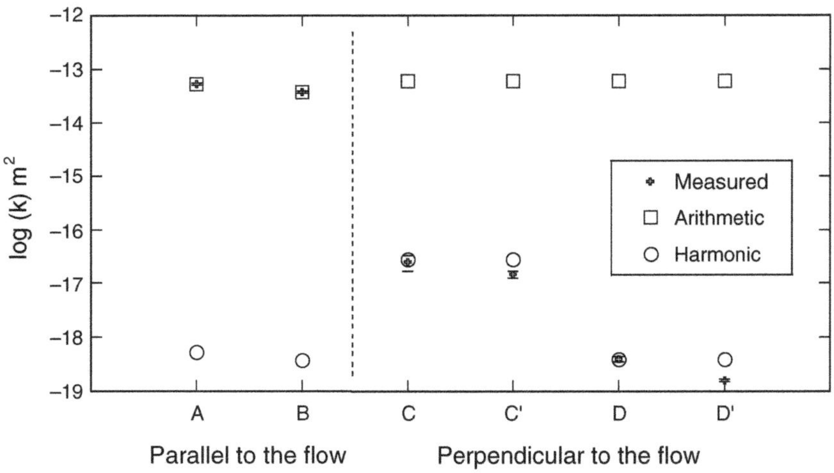


Figure 8
Permeability of mixed stacks in comparison to calculated arithmetic and harmonic averages of the permeability values of constituting sub-samples.

### *4.1. Effect of Sample Length and Percolation Theory*

Describing the geometrical features of a pore network is clearly a three-dimensional problem. The different lengths of intact samples vary the spatial scale only in the main direction of flow and the aspect ratio of sample length to radius $L/R$ covaries with length since the diameter is constant. The slender sub-samples of type A and B (Fig. 2) additionally vary the spatial scale perpendicular to the main flow direction. Introducing aspect ratio as a nondimensional scale parameter has obviously limited applicability because conduits are of finite size and therefore absolute scale matters.

For the sandstone variety with ~8% porosity, neither absolute size nor aspect ratio significantly affects permeability. In contrast, we observe a strong length-dependence of permeability for the variety with ~5% porosity (Fig. 4). In principle, an effect of sample length on the determined permeability may be caused by the approximate realization of the boundary conditions at the sample ends underlying the analytical solution of the one-dimensional diffusion equation employed to analyze our data (eqs. (2), (3)). ADACHI and DETOURNAY (1997) showed that a hydro-mechanical coupling has to be accounted for when sample ends are not subjected to the hydrostatic confining pressure, too. Within a continuum-mechanical approach, neglecting the coupling causes an apparent dependence of permeability on the aspect ratio of samples. The absence of a scale effect for the variety with ~8% porosity is not diagnostic because samples of this variety do not reach an aspect ratio of ~1 considered critical. The somewhat increased pressure dependence of the shortest samples (10 mm length) of the variety with ~5% porosity (Fig. 4a) may hint towards a change in boundary conditions with decreasing aspect ratio. However, the increase in permeability of the low porosity variety by orders of magnitude at a critical length (Fig. 9a) lets us prefer to interpret the observed size dependence of the low porosity variety as a percolation phenomenon.

By definition, the percolation threshold is the smallest porosity for which an interconnected cluster of pores exists with infinite size, the backbone (STAUFFER and AHARONY, 1994). At or above the percolation limit, the size dependence of sample properties is controlled by a characteristic length scale, the correlation length. For

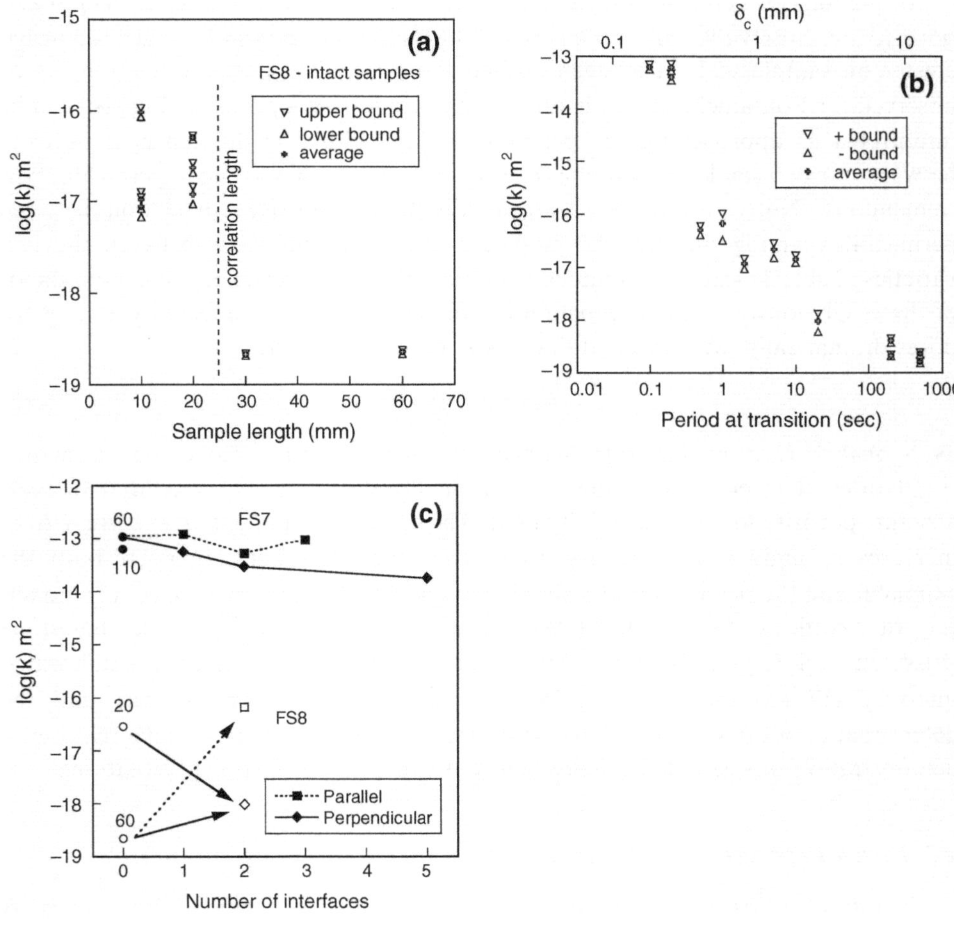


Figure 9
a) The effect of sample length on the long-period limit of permeability for the variety with ~5% porosity. b) Correlation between the long-period limit of permeability and the observed period at the transition to period-sensitive behavior for all tested samples. c) The effect of interfaces on the long-period limit of permeability for the two sandstone varieties (open symbols: FS8, ~5%; closed symbols: FS7, ~8%). Labels indicate sample length in mm.

systems with a size above the correlation length, transport properties become size-independent, i.e., a plateau of constant permeability exists. In systems with a size below the correlation length, clusters of less than infinite size may be large enough to substantially contribute to total flow in addition to the conduits forming the backbone. With increasing system size the amount of flow-relevant interconnected pore-volume not belonging to the backbone decreases and consequently permeability, too, but also its variability (see HUNT, 2005). While our observations regarding a critical length (Fig. 9a) are in good agreement with concepts from percolation theory, the small number of experiments clearly does not yield results that would withstand a check of statistical significance.

In the light of percolation theory, the dependence of permeability on length and porosity are closely related. Finite permeability is the very proof of an interconnected cluster on sample scale. The steep decline in permeability-porosity relationship as observed for Fontainebleau sandstone samples below ~8% porosity (Fig. 1b) can be interpreted as approaching the percolation threshold. The difference in porosity between our two sandstone varieties is just ~3%, yet the difference in permeability is tremendous. Note, relying on the generally found positive correlation between permeability and grain size, the moderate grain size difference between the two varieties (Table 1) leads to an underestimation of the difference of less than a factor of three. Obviously, a small reduction in porosity alters the connectivity among the pores dramatically. MARTYS and GARBOCZI (1992) suggested

$$k \propto (\phi - \phi_c)^n \tag{5}$$

as a scaled relation between permeability and porosity based on numerical simulations. The porosity at the percolation threshold, $\phi_c$, is introduced to scale absolute porosity to the connectivity limit. The apparent exponent, $n' \equiv \partial \ln k / \partial \ln \phi$, increases strongly when porosity approaches the percolation limit. If both the exponent and the percolation threshold are considered fit parameters, eq. (5) models the observations for Fontainebleau sandstone (Fig. 1b), synthetic limestone (FISCHER and PATERSON, 1992; ZHANG *et al.*, 1994), and texturally equilibrated quartzite (WARK and WATSON, 1998). Yet, our observations suggest that the determination of the connectivity limit from laboratory experiments requires a demonstration of size independence before it can be applied to field situations.

### *4.2. Period Dependence and Heterogeneity*

In our tests, oscillation period affects the determined permeability values. A transition from constant values to period-dependent permeability occurs at vastly different periods, ~500 s and ~0.1 s for the variety with ~5 (FS8) and ~8% (FS7) porosity, respectively. A similar dependence of permeability on the oscillation period was reported from previous laboratory experiments (e.g., BERNABÉ *et al.*, 2004; FISCHER and PATERSON, 1992). Four potential reasons for the period-dependence have been proposed: End effects, pressure sensitivity, inertia effects, and sample heterogeneity.

On reasons presented in the previous section, we do not follow up on the possibility of a significant period dependence related to effects at sample ends as suggested by ADACHI and DETOURNAY (1997). Pressure-sensitivity may affect the period dependence in our experiments. We find a tendency that a decrease in permeability with increasing period occurs for samples that exhibit a pronounced sensitivity to changes in effective pressure. The sigmoidal relation between permeability and oscillation period (Figs. 4b, 6b) is comparable to the results of network modeling when compressibility of either pore fluid or pore space are included (BERNABÉ, 1997).

The period dependence of permeability observed for the variety with ~8% porosity resembles the features associated with the concept of dynamic permeability. In particular, the slope of $T^{3/2}$ (Fig. 7) coincides with the theoretically expected slope (e.g., ZHOU and SHENG, 1989). However, three reasons support the notion that the partial agreement is coincidental and suggest that causes alternative to inertia effects should be considered for the bulk of our observations. Firstly, the critical skin depth of ~180 μm associated with the transition period exceeds our estimates of characteristic pore size by almost an order of magnitude (Table 1). Similar observations were made on Solnhofen limestone (FISCHER and PATERSON, 1992) for which the transition occurs at periods that are inconsistent with expected pore radii when the viscous skin depth scaling is used. The observed transition period of $\sim 10^4$ s corresponds to a viscous length scale $\delta_{SL} > 1\text{mm} \gg \bar{d}_{SL} \sim 5\,\mu\text{m}$, i.e., more than two orders of magnitude larger than the average grain size of Solnhofen limestone, $\bar{d}_{SL}$; furthermore, the slope depends on amplitude of the pore fluid oscillation (Fig. 10 of FISCHER and PATERSON, 1992). Secondly, our results exhibit an inverse relation between the plateau value and the transition period (Fig. 9b), in contrast to the dynamic scaling theories that to zero order predict a proportionality $k_0 \propto T_c$ (e.g., ZHOU and SHENG, 1989). Lastly, preliminary numerical analyses revealed that a complex-valued permeability according to the Johnson scaling concept (JOHNSON *et al.*, 1987) causes phase shift to increase when oscillation periods fall below the critical value, resulting in an apparent increase in permeability when evaluated with eqs. (2), (3). More work is needed to understand to what extent and how the used analysis of experiments reflects features of inertia-related dynamic permeability. We focus on the potential relation between period-dependence and heterogeneity in the remainder of this section.

The characteristic spatial scale of an oscillatory diffusion process is given by the penetration depth $r_p \sim \sqrt{DT}$ approximating the characteristic time scale for the period. For heterogeneity to significantly influence the result of an oscillatory test, the penetration depth had to be on the same scale for as the characteristic size of the heterogeneity that is necessarily smaller than the sample dimensions, i.e., $r_p \sim l_{het} < L$. To test the suggested relation between penetration depth and the oscillation period at which permeability becomes sensitive to it, we determine

$$r_{pc} \sim \sqrt{\frac{k_0 T_c}{\mu \phi \beta_f}}, \tag{6}$$

where we approximate the storage capacity $\beta_s \simeq \phi\beta_f$ because deviations from this approximation most likely do not reach an order of magnitude for our pressure-insensitive samples. For all but the intact samples of the variety with ~8% porosity, we find $r_{pc} \simeq L$ (Table 1) in general consistency with the interpretation that the onset of period dependence occurs when the penetration depth reaches the sample size. The intact samples with ~8% porosity (FS7) yield $r_{pc} \simeq 0.6\text{m} \gg L$, i.e., the transition to period-dependent behavior likely has a different origin than sample heterogeneity.

### *4.3. Effect of Interfaces and the Conduit Geometry*

The results for stacks of a single variety allow us to constrain geometrical properties of pores. Stacking of sub-samples can be addressed by two concepts. The interfaces can be considered as an interruption of the pore networks of the sub-samples or alternatively as separate hydraulic elements with specific transport properties. The dominance of either of these two effects depends on interface condition and orientation. Pressure dependence of permeability is not specifically diagnostic for either concept but an increase compared to intact samples demonstrates that some contribution to total flow comes from transport along the interfaces.

For the variety with ~5% porosity, the presence of parallel interfaces enhances permeability by more than two orders of magnitude (Fig. 9c). Assuming that the permeability is due to the transmissivity of the interfaces alone, the cubic law for flow between parallel plates gives an upper bound for the interface width of ~3 μm. The success of arithmetic averaging for mixed, flow-parallel stacks shows that the slender pieces of the variety with ~8% porosity (FS7) exhibit the same permeability as a full cylinder. Thus, in the more porous variety the directional interconnectivity of the network requires a scale lateral to main flow of less ~10 mm.

Perpendicular interfaces constitute a permeability reducing barrier (Figs. 5 and 6). Harmonic averaging of permeability of the sub-samples fails for stacks of discs of a single variety. The effect is larger for the variety with less porosity (Fig. 9c). Together with the above constraint on the hydraulic width of the interfaces, these observations suggest that the fraction of matched pore throats across the two touching surfaces controls the bulk permeability. Neglecting pore size distribution and spatial correlation for a crude estimate, the hypergeometric expectation value for two pore throats to match across the interface and to actually belong to the interconnected network of either sub-sample amounts to $E_{\text{match}} \sim N_{\text{pore}}\ \lambda_1 \lambda_2 \phi_1 \phi_2$. For this estimate, we assume a similar number $N_{\text{pore}}$ of pore throats on either side of the interface between sub-samples 1 and 2, and $\lambda$ denotes the probability for a given pore throat to belong to the interconnected network of a sub-sample. For the number of pore throats ending on an interface, $N_{\text{pore}}$, we found good agreement between a simple count of pores resolvable on the magnification scale of Fig. 1 and the conservative upper bound $N_{\text{pore}} \leq A_{\text{interface}}/\bar{A}_{\text{grain}} = (2R/\bar{d})^2$ (i.e., the number of pores per grain is identical to the grain's coordination number) using determined average grain diameters $\bar{d}$ (Table 1). Modeling the transport across the interface of a stacked sample of variety FS7 by identical tubes, tube radius $r_{\text{tube}}$ is limited by

$$r_{\text{tube}}^4 \leq \frac{8k_0 A_{\text{interface}}}{\pi N_{\text{pore}} \lambda^2 \phi^2} \leq \frac{2k_0 \bar{d}^2}{\lambda^2 \phi^2}. \tag{7}$$

The first bounding results from the length scale across which the macroscopic pressure difference $\Delta p$ acts, $\Delta p/L \leq \left|\vec{\nabla} p\right|_{\text{tube}} \leq \Delta p/\bar{d}$; the second results from the bound for $N_{\text{pore}}$. Assuming $\lambda_{\text{FS7}} \simeq 1$ based on the observed size-independence of permeability of FS7-samples, we determine $r_{\text{tube,FS7}} < 27$ µm, an estimate comparable with the critical hydraulic radius for intact samples derived including resistivity measurements (Table 1). The constraint $E_{\text{match,FS8}} \sim N_{\text{pore,FS8}} \lambda_{\text{FS8}}^2 \phi_{\text{FS8}}^2 > 1$ yields $\lambda_{\text{FS8}} > 0.1$ and thus $r_{\text{tube,FS8}} < 7.5$ µm. This tube radius estimate differs from the estimate for FS7 by only half an order of magnitude but exceeds the critical hydraulic radius (Table 1) by orders of magnitude highlighting the importance of interconnectivity in comparison to conduit size for the sample with ~5% porosity that is close to the percolation limit according to the permeability porosity relation (Fig. 1). A comparable conduit size for the two varieties is also in agreement with the microstructural quantities gained from SEM images (Table 1). Furthermore, the estimated hydraulic width of the interface falls below the estimated tube sizes yielding internal consistency among the estimates and explains the dual effect of interfaces on the variety with less porosity. The flow parallel interfaces dominate transport despite their small width since the natural network in an intact sample of 60 mm length corresponds to a single capillary with $r_{\text{tube,FS8}}$. In contrast, lateral transport on the narrow perpendicular interfaces is ineffective.

## 5. *Conclusions*

We investigated the hydraulic behavior of intact and stacked samples of two varieties of Fontainebleau sandstone with porosities of ~5 and ~8% whose permeability is insensitive to variation in effective pressure in accord with the observed ultrasonic velocities close to the upper Hashin-Shtrikman bound. The difference in permeability of six orders between the two varieties can be modeled by a relation $k \propto (\phi - \phi_c)^n$ motivated by percolation theory with a percolation threshold of $\phi_c$ ~5%. The pronounced length dependence of permeability of the variety with the lower porosity (~5%) suggests a correlation length of ~25 mm and provides further evidence for a network close to the percolation threshold. Results for stacked samples constrain geometrical properties of pores. We find characteristic sizes for the hydraulic conduits on the order of 10 μm. Only, for the variety with ~8% porosity, the critical radius according to KATZ and THOMPSON (1986) coincides with this estimate from the interface effects. The critical radius approach seems to overestimate the importance of conduit size compared to interconnectivity for networks approaching the percolation limit. Period-sensitivity of permeability seems to be affected by the length scale of heterogeneity in a sample and also by the fluid or frame compressibility. Numerical modeling will be important for further investigating and substantiating these relations, in particular for networks close to the percolation threshold.

## REFERENCES


ADACHI, J.I. and DETOURNAY, E. (1997), *A poroelastic solution of the oscillating pore pressure method to measure permeabilities of "tight" rocks,* Int. J. Rock Mech. Min. Sci and Geomech. Abstr. *34*(3/4), 430, 62.

ANDRADE, J.S., ALMEIDA, M.P., MENDES FILHO, J., HAVLIN, S., SUKI, B., and STANLEY, H.E. (1997), *Fluid flow through porous media: The role of stagnant zones,* Phys. Rev. Lett. *82*, 3901–3904.

BERNABÉ, Y. (1997), *The frequency dependence of harmonic fluid flow through networks of cracks and pores,* Pure Appl. Geophys. *149*, 489–506.

BERNABÉ, Y. and BRUDERER, C. (1998), *Effect of the variance of pore size distribution on the transport properties of heterogeneous networks,* J. Geophys. Res. *103*(B1), 513–525.

BERNABÉ, Y., MOK, U., and EVANS, B. (2006), *A note on the oscillating flow method for measuring rock permeability,* Int. J. Rock Mech. Min. Sci and Geomech. Abstr. *43*(2), 311–316.

BERNABÉ, Y., MOK, U., EVANS, B., and HERRMANN, F.J. (2004), *Permeability and storativity of binary mixtures of high- and low-permeability materials,* J. Geophys. Res. *109*, B12207, doi: 10.1029/2004JB003111.

BOURBIE, T. and ZINSZNER, B. (1985), *Hydraulic and acoustic properties as a function of porosity in Fontainebleau sandstone,* J. Geophys. Res. *90*, 11524–11532.

BRACE, W.F., WALSH, J.B., and FRANGOS, W.T. (1968), *Permeability of granite under high pressure,* J. Geophys. Res. *73*, 2225–2236.

CHARLAIX, E., KUSHNICK, A.P., and STOKES, J.P. (1988), *Experimental study of dynamic permeability in porous media,* Phys. Rev. Lett. *61*(14), 1595–1598.

DAVID, C. and DAROT, M., *Permeability and conductivity of sandstone,* In *Rock at Great Depth* (ed. Maury, D.) (A.A. Balkema, Rotterdam, 1989) pp. 203–209.

DOYEN, P.M. (1988), *Permeability, conductivity, and pore geometry of sandstone,* J. Geophys. Res. *93*, 7729–7740.

FISCHER, G.J. *The determination of permeability and storage capacity: Pore pressure oscillation method.* In *Fault Mechanics and Transport Properties of Rocks.* (eds. B. Evans, and T.-f. Wong) (Academic Press, San Diego, 1992) pp. 187–211.

FISCHER, G.J. and PATERSON, M.S. *Measurement of permeability and storage capacity in rocks during deformation at high temperature and pressure.* In *Fault Mechanics and Transport Properties of Rocks* (eds. B. Evans, and T.-f. Wong) (Academic Press, San Diego, 1992) pp. 213–252.

FREDRICH, J.T., GREAVES, K.H., and MARTIN, J.W. (1993), *Pore geometry and transport properties of Fontainebleau sandstone,* Int. J. Rock Mech. Min. Sci and Geomech. Abstr. 30, 691–697.

HASHIN, Z. and SHTRIKMAN, S. (1963), *A variational approach to the elastic behaviour of multiphase materials,* J. Mech. Phys. Solids *11*, 127–140.

HILFER, R., *Transport and relaxation phenomena in porous media.* In *Advances in Chemical Physics* (eds. I. Prigogine, and S.A. Rice), (John Wiley and Sons, New York, 1996) pp. 299–424.

HUNT, A.G., *Percolation Theory for Flow in Porous Media* (Springer, Berlin, 2005).

JOHNSON, D.L., KOPLIK, J., and DASHEN, R. (1987), *Theory of dynamic permeability and tortuosity in fluid-saturated porous media*, J. Fluid Mech. *176*, 379–402.

KATZ, A.J. and THOMPSON, A.H. (1986), *Quantitative prediction of permeability in porous rocks,* Phys. Rev. B, *34*, 8179–8181.

KRANZ, R.L., SALTZMAN, J.S., and BLACIC, J.D. (1990), *Hydraulic diffusivity measurements on laboratory rock samples using an oscillating pore pressure method,* Int. J. Rock Mech. Min. Sci and Geomech. Abstr. *27*(5), 345–352.

MARTYS, N. and GARBOCZI, E.J. (1992), *Length scales relating the fluid permeability and electrical conductivity in random two-dimensional model porous media,* Phys. Rev. B, *46*, 6080–6090.

MAVKO, G., MUKERJI, T., and DVORKIN, J. *The Rock Physics Handbook* (University Press, Cambridge, 1998).

PRIDE, S., *Relationships between seismic and hydrological properties*. In *Hydrogeophysics* (eds. Y. Rubin, and S.S. Hubbard) (Springer, Berlin, 2005) pp. 253–291.

RENARD, P. and DE MARSILY, G. (1997), *Calculating equivalent permeability: A review,* Adv. Wat. Resourc. *20*(5–6), 253–278.

SHENG, P. and Zhou, M.-Y. (1988), *Dynamic permeability in porous media,* Phys. Rev. Lett. *61*(14), 1591–1594.

SONG, I. and RENNER, J. (2006), *Linear pressurization method for determining hydraulic permeability and specific storage of a rock sample,* Geophys. J. Int. *164*(3), 685–696.

STAUFFER, D. and AHARONY, A. *Introduction to Percolation Theory* (Taylor and Francis, London, 1994).

STEWART, C.R., LUBINSKI, A., and BLENKARN, K.A. (1961), *The use of alternating flow to characterize porous media having storage pores,* Trans. AIME *222*, 383–389.

TORQUATO, S., *Random Heterogengeous Materials* (Springer, New York, 2002).

TURNER, G.A. (1958), *The flow-structure in packed beds,* Chem. Engng. Sci. *7*, 156–165.

TURNER, G.A. (1959), *The frequency response of some illustrative models of porous media,* Chem. Engng. Sci. *10*, 14–21.

WARK, D.A. and WATSON, E.B. (1998), *Grain-scale permeabilities of texturally equilibrated, monomineralic rocks,* Earth Planet. Sci. Lett. *164*(3–4), 591–605.

ZHANG, S., PATERSON, M.S., and COX, S.F. (1994), *Porosity and permeability evolution during hot isostatic pressing of calcite aggregates,* J. Geophys. Res. *99*(8), 15,741–15,760.

ZHOU, M.-Y., and SHENG, P. (1989), *First-principles calculations of dynamic permeability in porous media,* Phys. Rev. B, *39*(16), 12,027–12,039.

(Received April 28, 2005, revised March 23, 2006, accepted March 29, 2006)
Published Online First: September 14, 2006

Pure appl. geophys. 163 (2006) 2125–2140
0033–4553/06/102125–16
DOI 10.1007/s00024-006-0120-4



Pure and Applied Geophysics

# Determination of Porosity and Permeability out of Dispersion Analyses of Borehole Wall Displacements

Michael Albrecht,[1] and Vladimir Mansurov[2]

*Abstract*—Waves carried in the borehole can be measured by the use of controlled sources inside the borehole and receivers located in the same borehole as well. This article gives the theoretical background and develops a new method to determine rock-physical parameters out of such controlled measurements of waves carried inside the borehole. Theoretical dispersion curves of the group velocities are matched to dispersion analyses of registrations obtained in the borehole. This inversion process intends to determine the *in situ* porosity and permeability of the rock.

**Key words:** Porosity, permeability, borehole waves, dispersion, reservoir, borehole tool.

## *1. Introduction*

The determination of rock parameters is widely understood as a determination that takes place outside the rock's dominating physical environment. Samples are drawn, taken out of the original environment and conclusions are made in regards to the samples' original state.

The authors are especially interested in providing methods to the reservoir engineering industry that allow the *in situ* determination of rock parameters.

This paper gives the interested industry the analytical tools and the borehole tool design concept to establish a project that targets the *in situ* determination of reservoir porosity and permeability.

## *2. Transmission Properties of a Borehole*

If compression portions of seismic waves encounter a borehole, an elementary volume of this borehole filled with fluid will undergo a change in volume as shown in Figure 1.

[1] TBA Technology, 1696 E. 3300 S., Salt Lake City, Utah, 84106, U.S.A.

[2] Siberian Aerospace State University, "Krasnoyarskiy Rabochiy" Ave., Krasnoyarsk, 660014, Russia.

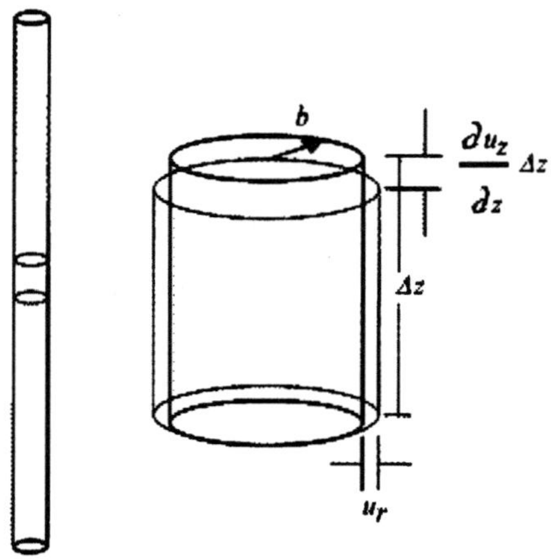


Figure 1
Deformation of an elementary volume filled with fluid in a borehole.

In the following it will be assumed that the considered wavelengths are significantly greater than the borehole diameter and that a quasi-static pressure change will arise because of the change in volume.

The pressure change $p(z,t) = P(z)e^{-i\omega t}$ in the fluid arising with the change in volume, as well as the displacement $u_z(z,t) = U(z)e^{-i\omega t}$ axial to the borehole, are functions of the borehole coordinate and of the time.

According to WHITE (1983) the following equation of motion can be established:

$$\left(\frac{\partial p}{\partial z} dz\right)(\pi \cdot b^2) = -\rho_{Fl}(\pi \cdot b^2 \cdot dz)\frac{\partial^2 u_z}{\partial t^2} \tag{1}$$

$b$ Borehole radius

$\rho_{Fl}$ Density of the borehole fluid

The pressure build-up within the fluid follows Equation (2).

$$P = -\mathrm{B}_{Fl}\left(\frac{dV}{V}\right) \tag{2}$$

$B_{Fl}$ Bulk modulus of the borehole fluid

$dV$ Change in volume

$V$ Starting volume of the elementary volume

During the pressure buildup changes in volume occur as shown in Figure 1, and they are dominated by a portion axial and radial to the borehole, following Equation (3).

$$dV = -V \cdot \frac{P}{B_{\mathrm{FL}}} = \left[\pi b^2 \left(\frac{\partial u_z}{\partial z}\right) dz + 2\pi b \cdot dz \cdot u_r\right]. \tag{3}$$

On the basis of $V = \pi * b^2 * dz$ this results in Equation (4).

$$\frac{P}{B_{\mathrm{Fl}}} = -\left(\frac{\partial u_z}{\partial z} + 2\frac{u_r}{b}\right). \tag{4}$$

Taking into consideration the input assumptions—low frequencies and quasi-static pressure changes—a linear equation between radial borehole displacement and pressure is developed, as shown in LAMB (1960), Equation (5).

$$u_r = \frac{b}{2M} \cdot P \tag{5}$$

$M$ Borehole parameter

The borehole parameter $M$ describes the borehole structure (piped, unpiped, type of piping) and the surrounding rock. Considering Equation (4) we obtain Equation (6).

$$\frac{\partial u_z}{\partial z} = -P\left(\frac{1}{B_{\mathrm{Fl}}} + \frac{1}{M}\right). \tag{6}$$

For the propagation of a wave in the impermeable borehole the following applies (LAMB 1960):

| | |
|---|---|
| $M = \mu + \frac{Eh}{2b}$ | piped borehole |
| h | Wall thickness |
| E | Young's modulus of the casing |
| $\mu$ | Shear rigidity of the rock |
| $M = \mu$ | unpiped borehole |

### 3. *Wave Equation for Axial Borehole Wall Displacements*

Combining EquationS (1), (4) and (6) result in the wave equation as shown in Equation (7).

$$\frac{\partial^2 u_z}{\partial z^2} = \frac{1}{c_T^2} \cdot \frac{\partial^2 u_z}{\partial t^2}. \tag{7}$$

The propagation velocity for the wave in the impermeable borehole is calculated according to Equation (8).

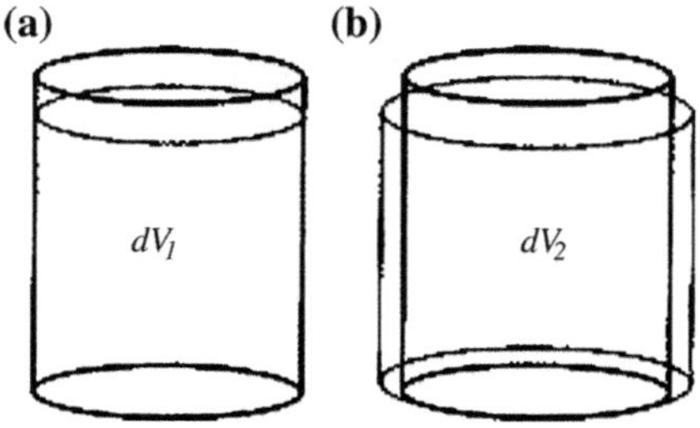


Figure 2
Development of waves carried in the borehole due to changes in volume caused by fluid compression (a) and expansion of the borehole wall (b).

$$c_T = \left[\rho_{Fl}\left(\frac{1}{B_{Fl}} + \frac{1}{M}\right)\right]^{\frac{1}{2}} \tag{8}$$

As can be seen from Equation (6), waves in the impermeable borehole are generated by volume changes $dV_1$ and $dV_2$ (Equations 9, 10 and 11), which are shown in Figures 2(a) and (b).

$$dV_1 = -\pi b^2 dz \frac{P}{B_{Fl}}, \tag{9}$$

$$dV_2 = -\pi b^2 dz \frac{P}{M}, \tag{10}$$

$$\frac{\partial u_z}{\partial z} = \frac{dV_1 + dV_2}{V}. \tag{11}$$

Contrary to that, according to WHITE (1983) the permeable borehole is characterized by a loss of fluid into the surrounding rock (see Figure 3), resulting in frequency-dependent volume changes, following Equation (12).

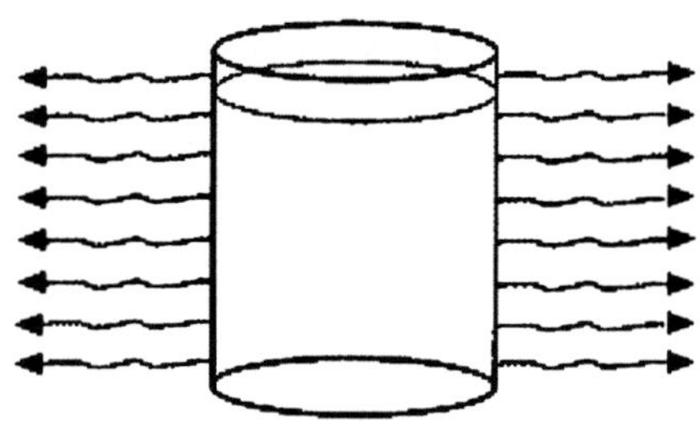

Figure 3
Development and influencing of waves carried in the borehole due to changes in volume caused by the inflow of fluid into the surrounding rock.

$$dV_3 = -2\pi b \cdot dz \frac{P}{i\omega Z(\omega)}. \tag{12}$$

$Z(\omega)$ is the borehole wall impedance that is equal to the ratio of pressure build-up and fluid velocity amplitude into the surrounding rock.

The wall impedance $Z(\omega)$ describes the ratio of pressure to average fluid-flow velocity of the borehole fluid into the surrounding rock and is calculated following Equations (13) and (14).

$$m = \frac{\phi\eta}{\kappa B_{Fl}} \tag{13}$$

$\phi$ Porosity of the rock

$\kappa$ Permeability of the rock

$\eta$ Viscosity of the borehole fluid

$$Z(\omega) = \frac{\eta}{\kappa} \frac{K_0(\sqrt{i\omega m} \cdot b)}{\sqrt{i\omega m} \cdot K_1(\sqrt{i\omega m} \cdot b)} \tag{14}$$

$K_0$ modified Bessel function of the 0th order

$K_1$ modified Bessel function of the 1st order

The modified Bessel functions will be resolved at this point not as an approximation, but as a general approach (BRONSTEIN and SEMENDJAJEW (1991), BRONSTEIN *et al.*, 1995).

All changes in volume mentioned above lead to a frequency-dependent absorption and propagation rate of the waves carried in the borehole:

$$\alpha_T(\omega) + \frac{i\omega}{c_T(\omega)} = i\omega\sqrt{\rho_{Fl}\left(\frac{1}{B_{Fl}} + \frac{1}{\mu} + \frac{2}{i\omega b Z(\omega)}\right)}. \tag{15}$$

To calculate the absorption coefficient $\alpha_T(\omega)$ and the phase velocity $c_T(\omega)$, Equation (15) is separated into a real and an imaginary part (see Equations 16 to 18).

$$i\omega\sqrt{\rho_{Fl}\left(\frac{1}{B_{Fl}} + \frac{1}{\mu} + \frac{2}{i\omega b Z(\omega)}\right)} = D_1 + iD_2 = D, \tag{16}$$

$$\alpha_T(\omega) = real\{D(\omega)\} = D_1(\omega), \tag{17}$$

$$c_t(\omega) = \frac{\omega}{img\{D(\omega)\}} = \frac{\omega}{D_2(\omega)}. \tag{18}$$

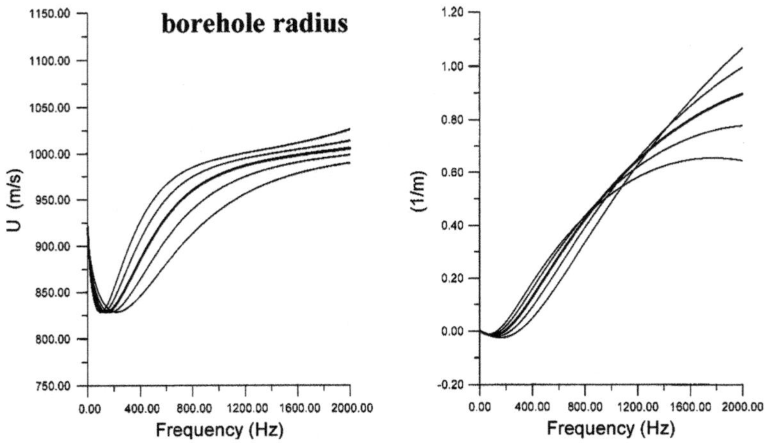


Figure 4
Group velocity and absorption coefficient as a function of the borehole radius (0.08–0.2 m). The lines in the middle belong to the borehole radius of 0.10 m.

## *4. Determination of Displacement and Group Velocity*

The standard solution for Equation (7) is the plane wave solution as a function of the spacing $\Delta z$:

$$u_z(z + \Delta z, t, \omega) = u_o \cdot e^{-i \cdot \omega t} e^{-\left[\alpha_T(\omega) + i\frac{\omega}{c_T(\omega)}\right] \cdot \Delta z}. \tag{19}$$

Figure 4 shows the absorption coefficient and the group velocity for boreholes of different radius. The group velocity $u_T$ is calculated according to Equation (21) out of the phase velocity and derives out of Equation (20) (MUELLER, 1990), with $k$ being the wave number.

$$u_T = \frac{d\omega}{dk} = c_T + k\frac{dc_T}{dk}, \tag{20}$$

Table 1

*Borehole parameters corresponding to the granite in Fenton Hill, New Mexico (FEHLER, 1982)*

| | | |
|---|---|---|
| Rock: | | |
| Porosity | 0.30 | |
| Permeability | $1000*10^{-15}$ | $m^2$ |
| Shear rigidity | $3.3*10^{10}$ | $kg/(m*s^2)$ |
| Borehole and pore fluid: | | |
| Viscosity | $1.4*10^{-4}$ | kg/(m*s) |
| Bulk modulus | $1.19*10^9$ | $kg/(m*s^2)$ |
| Density | $10^3$ | $kg/m^3$ |

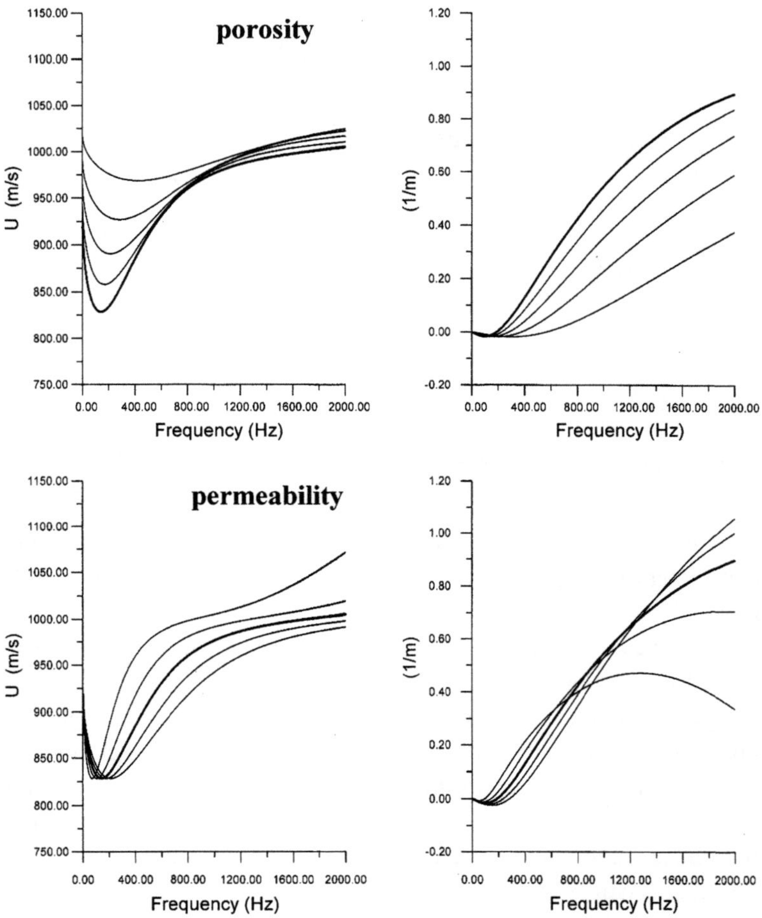


Figure 5

Group velocity and absorption coefficient as a function of porosity (0.10–0.30) and permeability ($500–1500 * 10^{-15}$ m$^2$). The top lines of the porosity dependency belong to a porosity of 0.30. The middle lines of the permeability dependency belong to a permeability of $1000 * 10^{-15}$ m$^2$.

$$u_t(\omega) = \frac{c_t(\omega)}{1 - \frac{\omega}{c_t(\omega)} \frac{dc_t(\omega)}{d\omega}}. \tag{21}$$

## *5. Modeling and Dispersion Analysis*

The parameters on which the presented calculations are based upon are shown in Table 1:

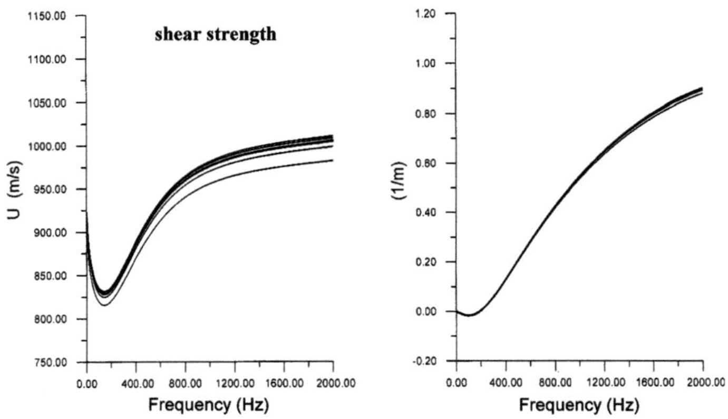


Figure 6
Group velocity and absorption coefficient as a function of shear rigidity ($1.3 - 5.3 * 10^{10}$ kg/(m $*$ s$^2$)). The lines in the middle belong to a shear rigidity of $3.3 * 10^{10}$ kg/(m $*$ s$^2$).

Figures 4 to 8 show both, the group velocity and the absorption coefficient, based on the parameters shown in Table 1. As reference the 0.10 m borehole radius is shown bold.

The dependency of the group velocity on the parameters can essentially be differentiated into two basic types; on the one hand there is a frequency shift of the airy phase with unchanged group velocity for variations of the borehole radius, the permeability and the viscosity, and on the other hand there is a change in the group velocity of the airy phase at constant frequency for variations of the bulk modulus and the density of the borehole fluid. The latter shows in addition an almost parallel shift of the phase velocity over the whole frequency spectrum.

From Figure 6 can be concluded that variations in the shear rigidity of the rock have only a slight influence on the shape of the dispersion curves. It is possible to distinguish between the bulk modulus and density of the fluid by the different, frequency-dependent absorption characteristics.

The dependencies of the dispersion curves on the borehole radius and the permeability are definitely comparable, nonetheless the borehole radius can always be assumed to be constant, given common calibre surveys. The calibre accordingly serves as a fixed boundary value.

Evaluation of the dispersion behavior of waves carried in the borehole therefore permits the inversion of the porosity and the permeability of the rock.

## *6. Stimulation and Logging of Waves Carried in the Borehole for the Inversion of Rock-physical Parameters*

In order to collect field data that allow the determination of porosity and permeability it is necessary to achieve an impulse stimulation on the source

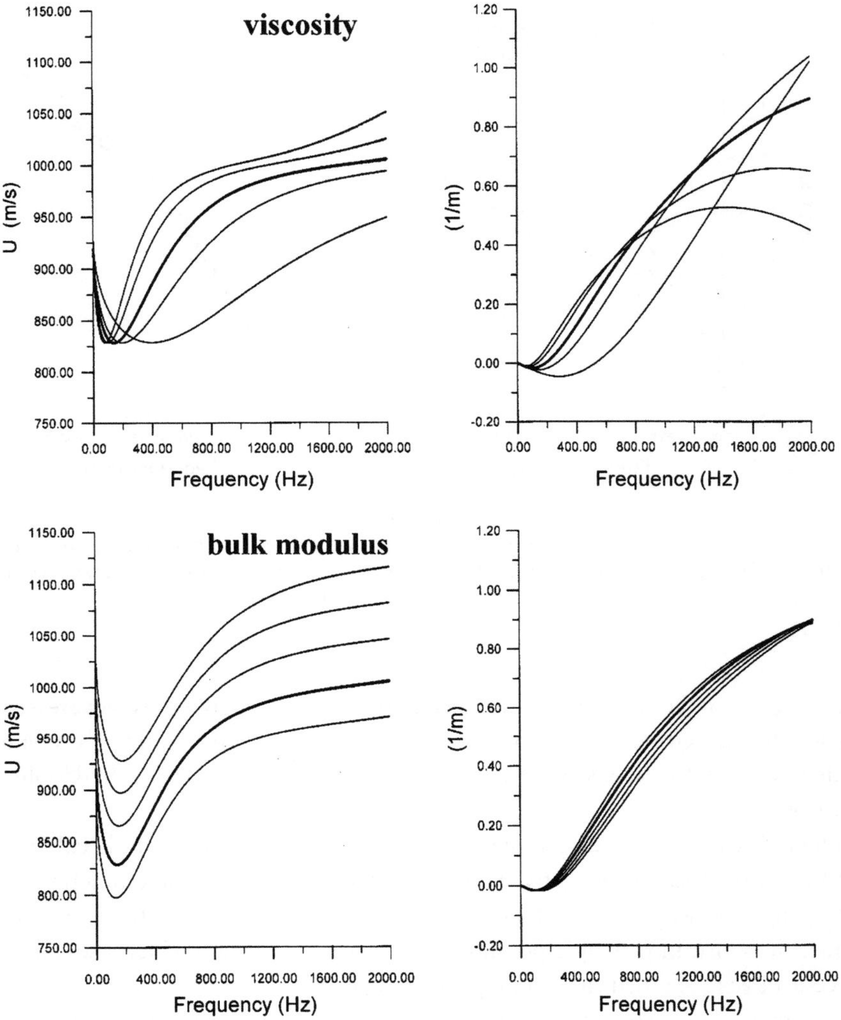


Figure 7
Group velocity and absorption coefficient as a function of viscosity ($0.5\text{–}2.5 * 10^{-4}$ kg/(m * s)) and bulk modulus ($1.1\text{–}1.5 * 10^{9}$ kg/(m * $s^2$)). The lines in the middle of the viscosity dependency belong to a viscosity of $1.4 * 10^{-4}$ kg/(m * s). The second line from the bottom of the bulk modulus dependency belongs to a bulk modulus of $1.19 * 10^{9}$ kg/(m * $s^2$).

side; a Dirac function would be optimal, or a sweep such as standard in vibroseismics.

Figure 9 shows a design concept of a measuring system which can be used to determine porosity and permeability. In order to avoid any misunderstandings in the basic design of this tool that targets the data acquisition of waves carried in the borehole, compared to regular seismic borehole tools. Figures 10 and 11 show more

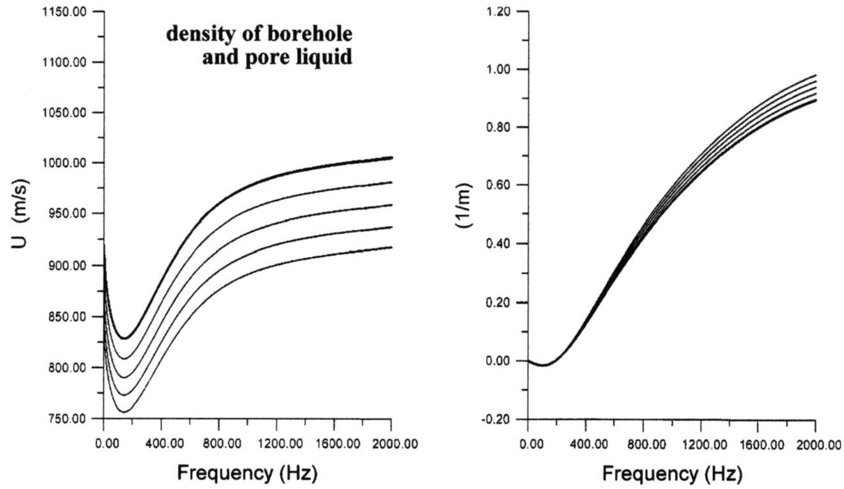


Figure 8
Group velocity and absorption coefficient as a function of the density of borehole and pore fluid (1.0 – $1.2 * 10^3$ kg/m$^3$). The group velocity line at the top represents a density of $10^3$ kg/m$^3$ that goes along with the absorption coefficient at the bottom.

details that can be considered as design proposals. A signal is generated in the borehole using a source (e.g., a pulser as shown in Figure 10), and it propagates over the distance $\Delta Z$ to the sensor module. The sensor incorporates a hydrophone (see Figure 11).

The influence of the connecting element (e.g., a standard borehole cable) will not be taken into consideration any further because of its negligible diameter compared to the generated wavelengths. Instead of a coupled system connected by a connecting element, it is also in principle possible to set up a free system which uses as a seismic source a pulser installed in an independent packer.

To invert porosity and permeability, dispersion curves from the dispersion analysis as calculated with Equation (22) or Equation (23) (see DZIEWONSKI *et al.*, 1969) have to be applied to the seismic registrations of the sensor module.
Impulse stimulation:

$$T_{(\omega)} = \Delta z / U_T(\omega) + T_0 \tag{22}$$

Sweep:

$$T_{(\omega)} = \Delta z / U_T(\omega) + T_0(\omega) \tag{23}$$

$T_o$ Point in time of impulse stimulation

$T_o(\omega)$ Phase position of stimulation frequency in the sweep

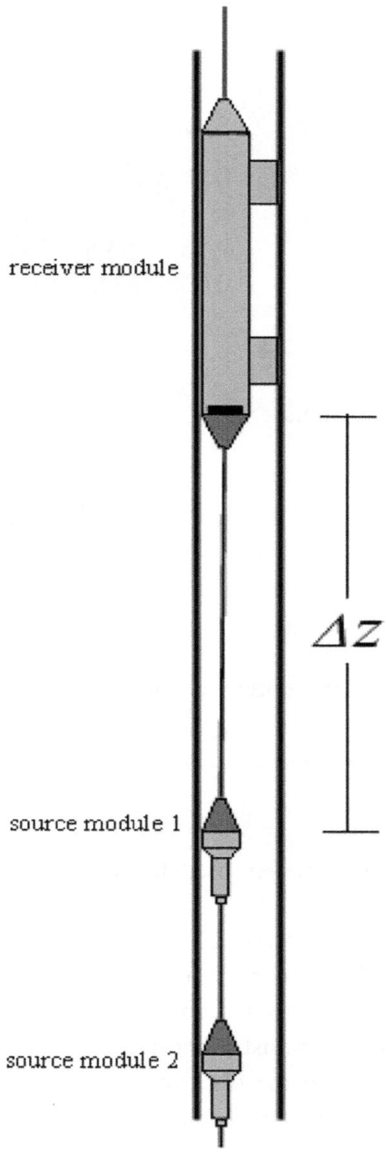


Figure 9
Logging of waves carried in the borehole. The source is located in a spacing $\Delta z$ below the sensor probe.

Figure 12 shows both, the source function and its synthetic seismic registration by the sensor module for different stimulation distances $\Delta Z$ and the frequency spectrum for a spacing of 20 m. The synthetic registration $U(Z + \Delta Z, t)$ is the result of the source signal $U_0(Z, t)$ according to Equation (18). The basis taken for

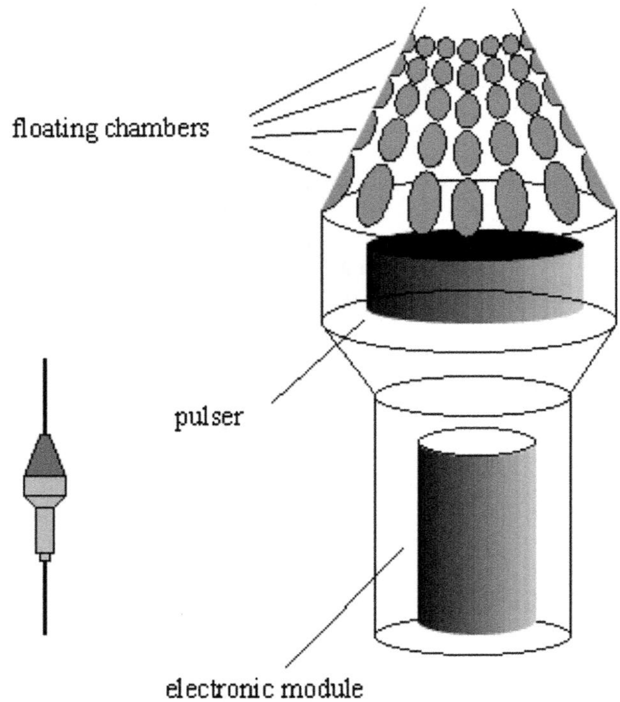


Figure 10
Structure of a pulser which can be used as a source (see Figure 9).

the calculations was the parameters used for Figure 4 with a borehole radius of 10 cm.

Figures 13 to 14 show the theoretical dispersion curves calculated out of Equation (21), together with the related, normalized dispersion analyses. For the synthetic registration, the lobes in the dispersion analysis are caused by the graphic software without any influence on demonstrating the relative process of matching the theoretical dispersion curves to the dispersion analysis of the seismic registration. The sensitivity of the adjustment with respect to the determination of the correct parameters is improved by an increase of the stimulation distance. While a correct parameter inversion is limited at a spacing of 10 m, a spacing of 20 m improves the inversion clearly. Any increase in the spacing results is an improvement of the inversion.

With the future development of suitable seismic sources, attention should be paid to ensuring that the principal energy of the source is in the frequency range of the airy phase. The airy phase with its flanks is the crucial criteria for the inversion of porosity and permeability.

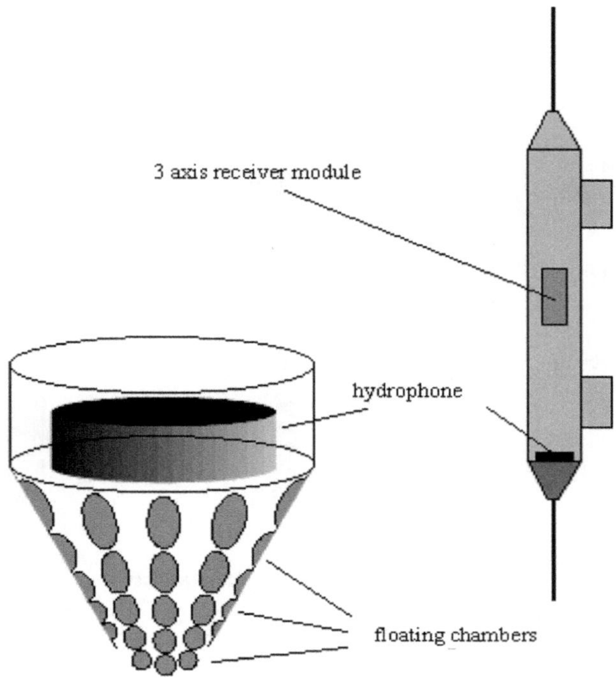


Figure 11
Sensor module for the inversion of waves carried in the borehole.

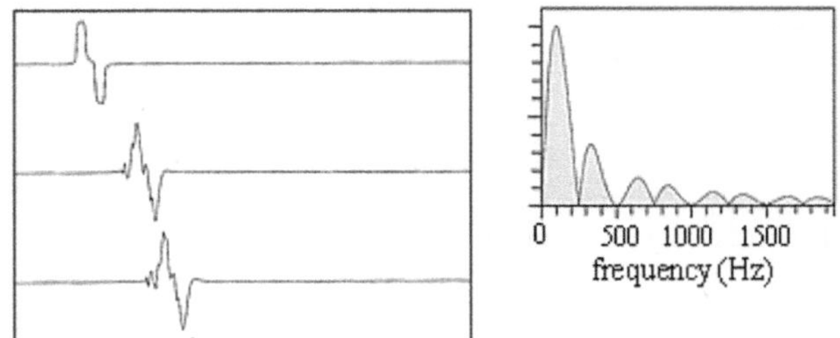


Figure 12
Source function and synthetic seismic registration of the sensor module for a spacing $\Delta z$ of 10 m, 15 m and 20 m and the frequency spectrum for the spacing $\Delta z$ of 20 m.

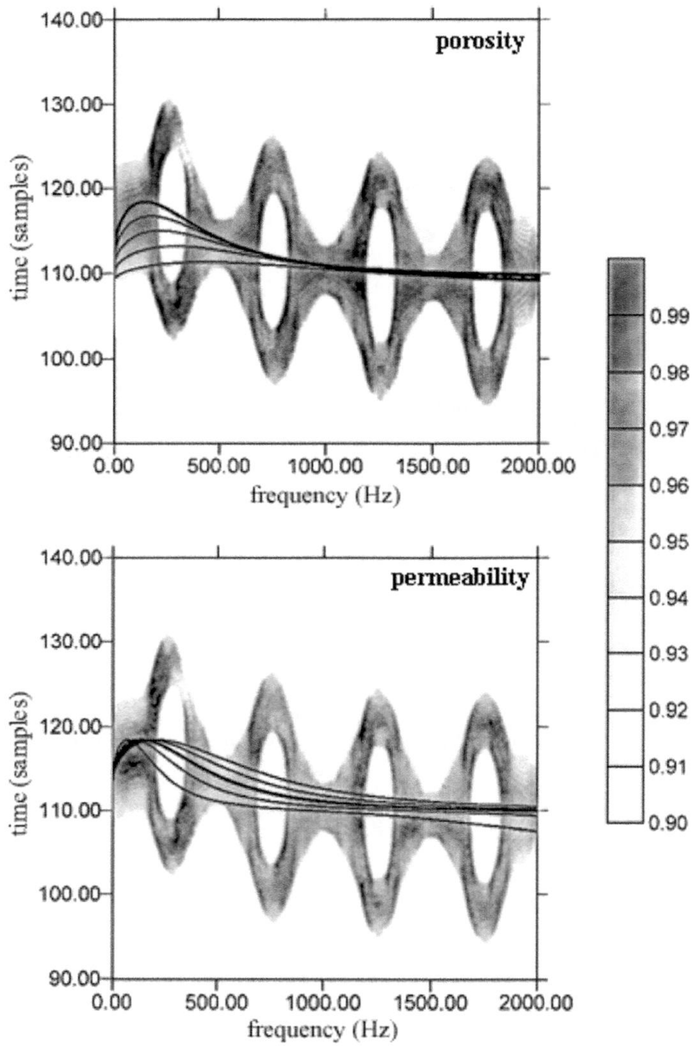


Figure 13
Normalized dispersion analysis of the seismic registration of the sensor module with the theoretical dispersion curves for porosity (a) and permeability (b) with a spacing $\Delta z$ of 10 m.

## 7. *Conclusion*

A coupled measuring system, containing a source and a receiver, can measure waves carried in the borehole that allow to determine the *in situ* porosity and the *in situ* permeability of the surrounding rock through a group velocity inversion process. The reservoir engineering industry is challenged to adopt this method and concept and pursue its field testing and implementation.

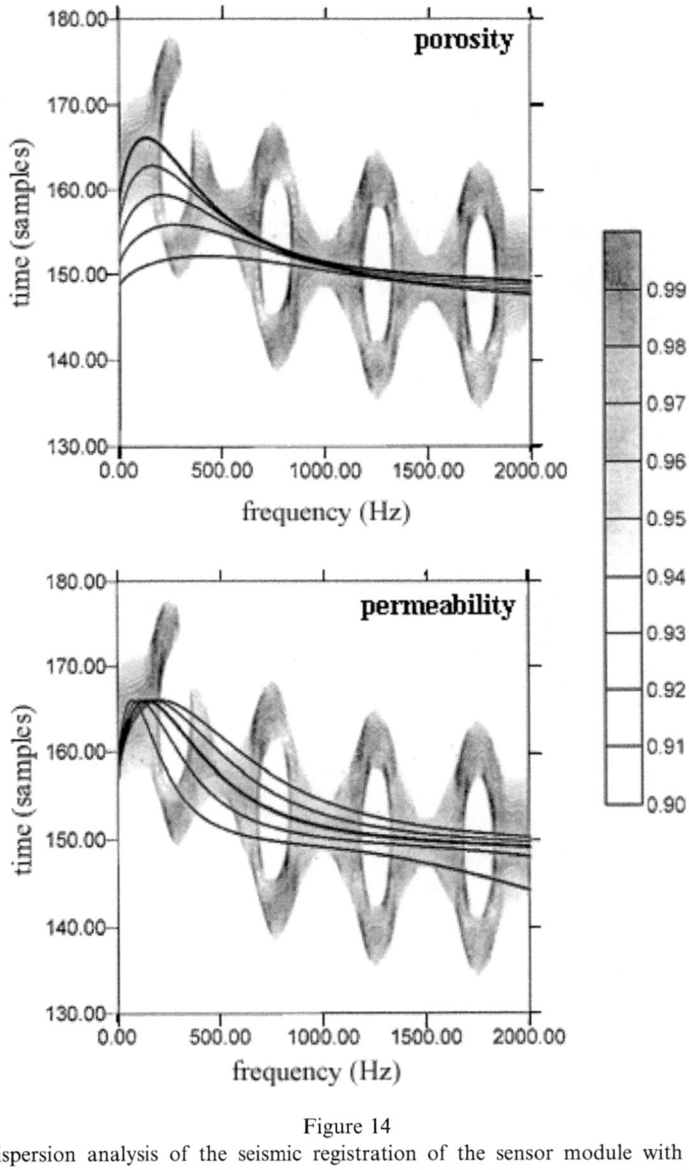


Figure 14
Normalized dispersion analysis of the seismic registration of the sensor module with the theoretical dispersion curves for porosity (a) and permeability (b) with a spacing $\Delta z$ of 20 m.

## References

Bronstein, I. N. and Semendjajew, K. A., *Taschenbuch der Mathematik* (Teubner Verlagsgesellschaft, Stuttgart/Leipzig, Verlag Nauka, Moskau 1991) ISBN 3-8154-2000-8, pp. 441.

Bronstein, I. N., and Semendjajew, K. A., Musiol, G., and Mühlig H., *Taschenbuch der Mathematik* (Verlag Harri Deutsch, Frankfurt am Main 1995), ISBN 3-8171-2002-8, pp. 399.

Dziewonski, A. M., Bloch, S., and Landisman, M., *A technique for the analyses of transient seismic signals* BSSA *59*, 427–444.

FEHLER, M., (1982) *Interaction of seismic waves with a viscous liquid layer*, BSSA *72*, 1, 55–72.
LAMB, H., *Statics, Including Hydrostatics and the elements of the Theory of Elasticity*, 3rd Edition, Cambridge University Press, 1960).
MUELLER, G., *Theorie elastischer Wellen*, Vorlesungsskript des Institut für Meteorologie und Geophysik, Universität Frankfurt, pp. 155, 1990.
WHITE, J. E., UNDERGROUND SOUND – APPLICATION OF SEISMIC WAVES (Elsevier Science Publishers, 1983 Amsterdam), ISBN 0-444-42139-4.

(Received November 8, 2005, revised January 30, 2006, accepted February 1, 2006)

Pure appl. geophys. 163 (2006) 2141–2152
0033–4553/06/102141–12
DOI 10.1007/s00024-006-0122-2



Pure and Applied Geophysics

# Fluid Pressure Variation in a Sedimentary Geothermal Reservoir in the North German Basin: Case Study Groß Schönebeck

ERNST HUENGES, UTE TRAUTWEIN, BJÖRN LEGARTH, and GÜNTER ZIMMERMANN

*Abstract*—The Rotliegend of the North German basin is the target reservoir of an interdisciplinary investigation program to develop a technology for the generation of geothermal electricity from low-enthalpy reservoirs. An *in situ* downhole laboratory was established in the 4.3 km deep well Groβ Schönebeck with the purpose of developing appropriate stimulation methods to increase permeability of deep aquifers by enhancing or creating secondary porosity and flow paths. The goal is to learn how to enhance the inflow performance of a well from a variety of rock types in low permeable geothermal reservoirs. A change in effective stress due to fluid pressure was observed to be one of the key parameters influencing flow properties both downhole and in laboratory experiments on reservoir rocks. Fluid pressure variation was induced using proppant-gel-frac techniques as well as waterfrac techniques in several different new experiments in the borehole. A pressure step test indicates generation and extension of multiple fractures with closure pressures between 6 and 8.4 MPa above formation pressure. In a 24-hour production test 859 $m^3$ water was produced from depth indicating an increase of productivity in comparison with former tests. Different depth sections and transmissibility values were observed in the borehole depending on fluid pressure. In addition, laboratory experiments were performed on core samples from the sandstone reservoir under uniaxial strain conditions, i.e., no lateral strain, constant axial load. The experiments on the borehole and the laboratory scale were realized on the same rock types under comparable stress conditions with similar pore pressure variations. Nevertheless, stress dependences of permeability are not easy to compare from scale to scale. Laboratory investigations reflect permeability variations due to microstructural heterogeneities and the behavior in the borehole is dominated by the generation of connections to large-scale structural patterns.

## *Introduction*

Increasing demand on renewable energy lead to the utilization of geothermal energy from areas with standard geothermal gradients as found in West and Central Europe. It becomes necessary to enhance energy recovery from depth. Therefore, two basic concepts were developed. The procedures distinguish between water bearing reservoirs and dry rocks and differ in the final operation scheme

- either creating an artificial heat exchanger at depth and using surface water for heat extraction from mostly dry rocks, e.g., Soultz-sous-Forêts (BAUMGÄRTNER *et al.*, 2004)

GeoForschungsZentrum Potsdam, Telegrafenberg, D-14473 Potsdam, Germany.
E-mail: huenges@gfz-potsdam.de

- or creating artificial pathways at depth to enhance the water flow from water bearing reservoir rocks, operating with a downhole pump installed in a production well and reinjection well, e.g. Groß Schönebeck (HUENGES *et al.*, 2004).

Both concepts are based on hydraulic fracturing techniques using variations in fluid pressure to design the reservoir. A review of several hydraulic fracturing methods is given by ECONOMIDES and NOLTE (1989). The variance of fluid pressure is upper-limited by the hydrofracturing itself, but this limit may not be reached if fluid escapes without failure (YARDLEY, 2001). Fracturing is a result of the interaction between external stresses and pore fluid pressure. Stresses are transmitted through sediment depending on its rock mechanical properties (PARK *et al.*, 2001). The same authors pointed out that fluid pressure reflects sediment deformation and compaction histories, as well as associated porosity/permeability changes and the presence of hydrocarbons. Thus, characterizing the distribution and the conditions for fracturing requires an understanding not only of the tectonic history, but also how the sediment evolved in geologic times. It is worth noting in this context the finding of BOURNE (2003) that natural fracture systems are often influenced by the presence of different layers with contrasting rock mechanical properties. In this case, tensile stress arises without the requirement for an additional body force such as internal fluid pressure and will concentrate within either the stiffest or softest layers according to the relative amount of layer-parallel versus layer-perpendicular compression. Beside this, production impairment from a promising reservoir has long been associated with formation damage due to infiltration of drilling fluids (GHALAMBOR and ECONOMIDES, 2002). An investigation into the effect of fluid pressure at depth must consider these aspects of the history of the reservoir rocks as well as the mechanical interaction of neighboring rock packages and the chemical interactions of fluids with the reservoir rocks.

While all these factors are important, the flow properties in the reservoir depend mainly on fluid pressure. The dependence of permeability on fluid pressure needs to be understood and quantified in order to scale permeability from the laboratory to natural conditions. Many authors ask for such experimental data (e.g., GUEGUEN *et al.*, 1996). The objective of this paper is to address the effects of fluid pressure during *in situ* experiments on reservoir scale as well as on core samples from the reservoir investigated at simulated *in situ* conditions.

Our study was performed in the former gas exploratory well Groß Schönebeck, which was reopened and deepened to 4294 m depth to serve as a geothermal *in situ* laboratory starting in December, 2000. The purpose of the down-hole laboratory was the development of a technology to increase permeability of deep aquifers using methods of hydraulic fracturing. The long-term goal is the ability to control the stimulation of a variety of rocks so that geothermal energy can be exploited from any kind of reservoir where it is needed. Conditions imposed on geologic formations suitable for our studies were: (1) Temperature above 120°C, which implied a formation at depths greater than 3000 m; (2) large regional extent so that results from this project may be extrapolated to other similar areas, and (3) a variety of lithologies available for investigation. The Rotliegend formation fulfills these requirements. It is a well-known formation because it has been extensively drilled for gas exploration and production. The Rotliegend rocks comprise clastic siltstones and sandstone layers generally overlying extrusive volcanic rock. Nine months after reopening the well a temperature of 149°C was measured at 4285.13 m depth. The formation pressure was determined using pressure logs after long-term fluid level observations reaching levels close to equilibrium 44.9 ± 0.3 MPa at 4220 m depth. The stimulation operations were realized and they created secondary flow paths and improved the inflow performance of the well (HUENGES *et al.*, 2004; ZIMMERMANN *et al.*, 2005a). In addition new data should help to explain the role of fluid pressure in the borehole scale.

TRAUTWEIN and HUENGES (2005) performed laboratory investigations on core samples from the same well. Changes of pore pressure under uniaxial strain boundary conditions induce a change of effective stress in these cores and this results in deformation of the rock samples. This deformation affects all macroscopic properties that were investigated: porosity, permeability, formation resistivity factor, compressional and shear-wave velocities.

The main purpose of this paper is to address this combined approach and to confirm the major role of fluid pressure in controlling reservoir properties.

## *Rotliegend Sandstones under Uniaxial Strain*

The laboratory experiments were designed to simulate reservoir depletion and injection of fluids in the Rotliegend reservoir sandstones (TRAUTWEIN and HUENGES, 2005). Pore pressure dependence of petrophysical properties was measured under uniaxial strain boundary conditions, i.e., zero lateral strain at constant overburden pressure. The applied level of *in situ* stress conditions on the reservoir rock samples was derived using methods recently compiled, for example, by ZOBACK *et al.* (2003). The axial stress of 100 MPa was maintained during the experiments. The initial confining pressure was 53 MPa according to the minimum horizontal *in situ* stress. Pore pressure was varied from 45 to 5 MPa enforcing uniaxial strain. This deformation procedure was chosen to produce a strain-dependent stress path close to reality. A change in fluid pressure cannot result in lateral deformation of the reservoir because of the confinement of the rock. Permeability, formation resistivity factor, and compressional as well as shear-wave velocities were recorded simultaneously and continuously during the deformation in the direction of maximum principal stress.

The results show that the stress development varies for different sandstones depending on the efficiency of pore pressure (i.e., Biot-factor). The stress anisotropy increases with decreasing pore pressure leading to stress-induced anisotropy of pore structure with preferred closure of horizontally oriented pore space, i.e., perpendicular to the axial load. Permeability incline of initially lower permeable sandstones evidences the opening of axially oriented pores at lower pore pressure (Fig. 1).

Figure 1 shows the permeability change based on its *in situ* values versus pore pressure. Permeability decreases with pore pressure due to the increase of effective stress, anisotropy of stress, which results from the pore pressure reduction under uniaxial strain condition. Permeability incline occurs for lower permeable sandstones at low pore pressure corresponding to the high anisotropy of stress, which results from the pore pressure reduction under uniaxial strain condition. Further investigations have shown that this behavior is mostly reversible with increasing pore pressure and repeatable for subsequent loading cycles. This pore pressure induced permeability increase does not correlate with deformation and porosity. Even at low pore pressure the observed porosity still decreases while permeability increases. Enhanced permeability at high differential stress for low porosity rocks is well documented in the literature (FAULKNER and RUTTER, 2003; HEILAND, 2003) but is usually accompanied by macroscopic measurable dilation seen in increasing volume before failure. The observed increase in permeability at constant axial compaction with impeded lateral deformation is rather unexpected. Obviously, any explanation for these observations has to include microstructural aspects.

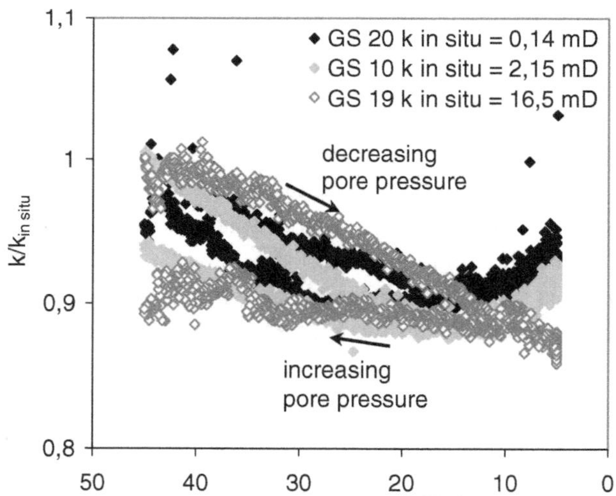


Figure 1
Permeability variations induced by pore pressure changes measured under uniaxial strain conditions on Rotliegend sandstone samples of the well Groß Schönebeck 3/90. Permeability is measured in axial direction, which is the direction of maximum stress ($\sigma_1$ = 100 MPa). Confining pressure varies with pore pressure to avoid lateral deformation. Permeability is normalized by its values determined at *in situ* stress conditions. The detectable accuracy is ±1%. (TRAUTWEIN and HUENGES, 2005).

### *Hydraulic History of the Well Groß Schönebeck 3/90 Including Stimulation Experiments*

The experiments in the deep are necessary to investigate the effect of fluid pressure on rocks comparable to those of the laboratory experiments under reservoir conditions. While the fluid-pressure investigations in the laboratory were performed under well-controlled conditions, the borehole tests are controlled by the heterogeneity of the reservoir.

After reopening and deepening the well Groß Schönebeck a production test was performed. During the 12 h test 167 $m^3$ water were extracted from the well at an average rate of 13.5 $m^3/h$ and a maximum pressure drawdown of 14 MPa at the production horizon. The flow logs show a significant inflow only between 4224 m and 4248 m, i.e., from the volcanic rocks. The inflow probably originates from single fracture zones. The Rotliegend sandstones with an expected permeability of 10–100 mD were completely blocked.

The frac operations in 2002 were designed to enhance the inflow performance by connecting the well to productive reservoir zones. Two consecutive zones within the Rotliegend sandstones were selected for stimulation (LEGARTH *et al.*, 2003). Here, core measurements show the most promising petrophysical reservoir properties, e.g., permeability, with respect to a productivity increase. The stimulation treatments were performed as hydraulic proppant fracturing operations. Proppants were used to support the fractures and to guarantee a long-term fracture aperture. The treatment intervals are located in the open-hole section of the well at depths between 4080 m and 4190 m and at temperatures of about 140°C. The technical demands were due to depth and temperature unprecedented in these challenging conditions. An open-hole-packer at the top and a sand plug at the bottom of each interval were used as hydraulic barriers. Applying this configuration the intervals were fracture-treated by injecting about 11 tons of proppant (high-strength ceramic grains) and over 200 $m^3$ of frac fluid (highly viscous gel) into the formation. The fracture treatments were conducted by means of two successive operations in each interval: First, a diagnostic treatment (datafrac), followed by the main treatment (mainfrac) with the proppant stages. The injection rate reached 150 $m^3/h$ at 16 MPa above formation pressure. The fracture closure pressure was determined by shut-in pressure decline analysis reaching a value of 8.4 MPa above formation pore pressure in the lower interval (LEGARTH *et al.*, 2003). Propped fractures were created in both intervals and the inflow behavior of the reservoir shortly after the operation—still at high fluid pressure in the reservoir—was decisively enhanced. However, LEGARTH *et al.* (2003) showed that the stimulation ratio, i.e., ratio productivity before and after the treatment, predicted by modeling could not be achieved. There were numerous reasons for this mismatch. Probably chemical and mechanical processes during closure after pressure release led to a reduced fracture conductivity.

To determine the low fluid pressure productivity after the first treatment, a moderate pumping test was carried out in the summer of 2002. The long-term production of 580 $m^3$ formation fluids within 37 days from 4200 m depth achieved steady-state conditions, i.e., constant flow rate and pressure. A depression of 0.8 MPa was observed at the end of the test. A significant change of flow parameters due to the previous stimulations in January/February, 2002 was not observed. Therefore, the proppant stimulation affected enhanced recovery at the high pressure state but no significant changes after pressure release.

Further stimulation of the reservoir rocks in this well was performed in two experiments in the winter of 2002/2003 and the fall of 2003 using the waterfrac technique. 11575 $m^3$ conditioned water was injected in different pressure steps (Fig. 2) with flow rates up to 80 $l\ s^{-1}$. Decreasing slopes of pressure versus injection rate were observed at injection flow rates above 9 $l\ s^{-1}$. The reduction of injection rate from 25 $l\ s^{-1}$ to 9 $l\ s^{-1}$ led after about 10 minutes to constant pressure interpreted as an equilibrium between fracture opening and draining of injection fluid into the formation (Fig. 3). The fracture closure pressure was determined at 6.4 MPa above formation pressure from the data of this equilibrium test and by pressure decline analysis of the subsequent shut-in phase. The fracture closure pressure always represents a *global* value determined from large-scale fracturing, valid for the fractured zone, where a significant net fracturing pressure share has to be attributable. The net fracturing pressure results from frictional pressure losses in the near-wellbore region and inside of the fracture. Therefore the fracture closure pressure cannot be directly compared with individual values of $\sigma_h$ min (*local* value)

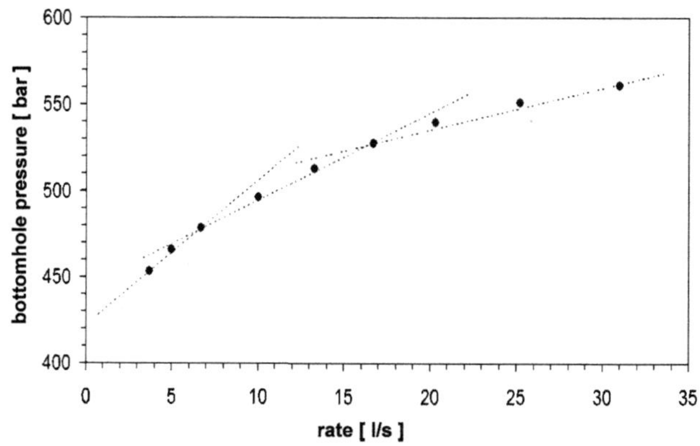


Figure 2

Pressure step test in the well Groß Schönebeck 3/90 (November 2003). The injection rate of slightly acid water (pH5) was kept constant at each step until the pressure measured at 4130 m depth approached equilibrium after about 10 minutes. The dotted lines indicate the decreasing slope with increasing injection rate interpreted as increasing mechanical behaviour of the system following the methods described by ECONOMIDES and NOLTE, (1989).

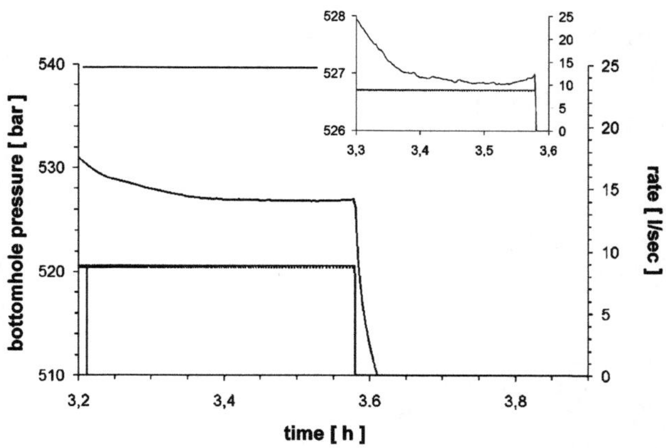


Figure 3
Equilibrium test to determine the fracture opening respectively closure pressure (see Fig. 2). This test was performed after reduction of injection rate (Fig. 2) from 25 l s$^{-1}$. An equilibrium between fracture opening and leak off is observed at 52.7 MPa at 4130 m. The fracture closure pressure was determined at 49.8 MPa.

determined via small-scale microfracturing (ECONOMIDES and NOLTE, 1989) or laboratory data. It is rather the average of the minimum principal stress of the zone covered by the created fracture system. Recent data of a production test shortly after the massive injection show a productivity index of 14 m$^3$h$^{-1}$MPa$^{-1}$ at pressure corresponding to open fracture. This productivity index seems to be sufficient for geothermal power production.

Temperature measurements before and after the last massive water injection showed that the massive injection test cooled down the lower section of the hole (II) (Fig. 4). There seems to be a seal between sections I and II at least close to the well. However, it cannot be decided from all recovered data which fracture system corresponds to the 6 MPa closure pressure, although section II is the most probable candidate. The significant changes of flow properties as the result of the hydraulic fracturing procedure occurred in section II.

A long-term injection test performed in December 2004 represents the final test of the well. A low injection rate of 2 l s$^{-1}$ is chosen to keep the system below the fracture opening pressure. After 18 days of injection (fresh water with pH5) a differential pressure of 3.7 MPa above formation pressure was observed. A shut-in of several months follows the injection phase. A final temperature measurement indicating the depth section of infiltration of the test water will be performed at the end of the test. The pressure transient data observed during injection and during shut-in follows a strong time$^{1/4}$ behavior indicating that a finite conductivity vertical fracture system is reacting (using the method of CINCO-LEY and SAMANIEGO, 1981). We can conclude that a low conductivity fracture system is open but with respect to the value of 3.7 MPa at the low flow rate of 2 l s$^{-1}$.

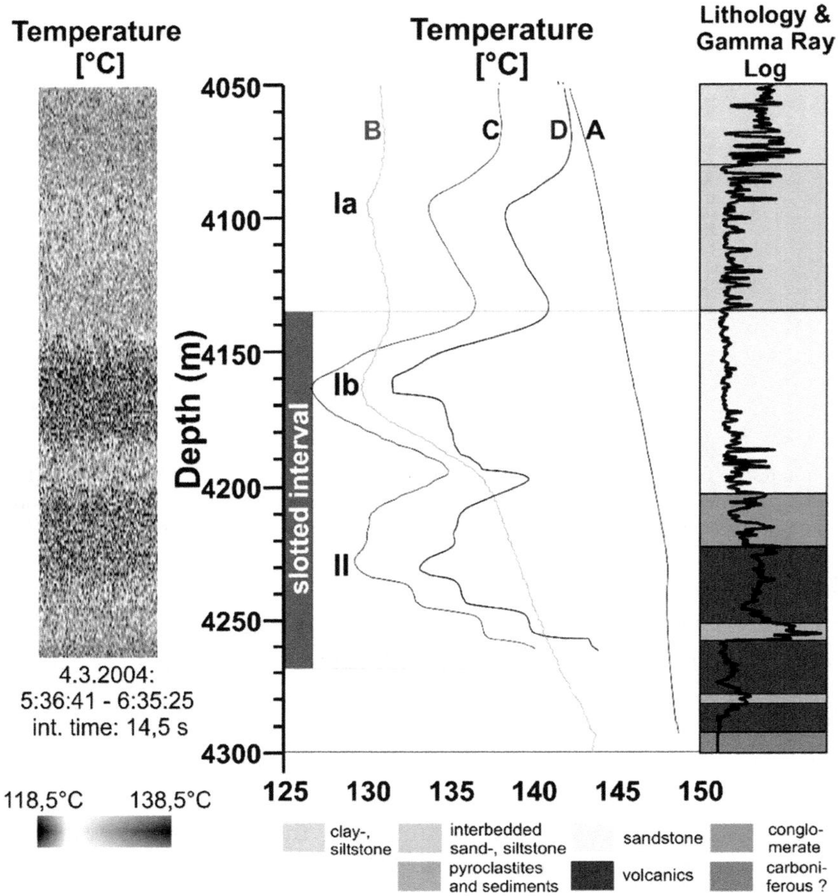


Figure 4
Temperature measurements after coldwater injection compared to recovered data before stimulation (HENNINGES *et al.*, 2005). Left: Profile using distributed temperature sensors. Middle: Temperature, A: after reopening of the well 2001, B: before massive stimulation 4. November 2003, C: 5. March 2004 after massive stimulation and D: 28. April 2004, 54 days later; hydraulic active zones indicated with Ia, Ib, and II. Slotted interval indicates the casing section with a slotted liner, other depth sections are cased. Right: Lithology and Gamma Ray.

## *Discussion*

Aspects (1) to (4) below must be considered to understand the measured *in situ* data of Groß Schönebeck: (1) The history of the experiments is important. The starting status of hydraulic properties was characterized by a manner of formation damage. There is no other explanation for the lacking inflow from the sandstones during the first test. (2) Significant changes of flow properties were observed as the result of the hydraulic fracturing procedure. (3) Flow properties depend on fluid pressure. Following the arguments by ECONOMIDES and NOLTE (1989) we can distinguish between the high-pressure case, above 6 MPa over formation pressure (mechanical reaction) and a low-pressure case (hydraulic reaction) beneath this value. The data can also support the findings of GUEGUEN *et al.* (1991) who concluded that crack networks in the Earth's crust respond in the following sequence: Stress-induced increase of fluid pressure produces hydraulic fracturing with the result that a connected hydraulic network is generated. Thereafter, due to fluid pressure release, the process of fracturing finishes. The case study Groß Schönebeck matches together with the above-described sequence, whereas the fluid pressure increase is due to hydraulic stimulation. The onset of the leak-off of fluids is due to the connection to pre-existing fractures or to high permeable rock layers and results in fluid pressure drop and termination of hydraulic fracturing. As a result a fracture network consisting of generated artificial fractures and pre-existing fractures close to the percolation threshold is generated. The observation of permeable fractured rock at the percolation threshold is consistent with findings of ZIMMERMANN *et al.* (2003, 2005a) for critically stressed crystalline and fractured rock. The experimental approach foresees injection induced varying pore pressure to change an existing stress field. Figure 2 shows several slope sections when observing the pressure as a result of injection into the formation of different horizons (e.g., sandstone, volcanic rocks see Fig. 4). We interpret the varying slopes in Figure 2 as indicators for successive opened fracture systems and the intersection of the slope with the measuring curve as the threshold for the fracture opening. One threshold is observed in Groß Schönebeck at an injection rate of 9 l $s^{-1}$, however, other thresholds are observed at higher injection rates. Therefore, the heterogeneity of the reservoir becomes an important issue. (4) We observed inflow into the reservoir at different injection pressures, e.g., higher pressure values by injecting into the probably damaged sandstones at the first fracturing treatment with about 8.4 MPa above formation pressure and 6.4 MPa in zone II during the waterfrac (Fig. 4).

Transport properties of the same rocks at the laboratory scale are functions of the microstructure (TRAUTWEIN and HUENGES, 2005). For different types of sandstones varying fluid pressure dependences of permeability were measured. At comparable fluid pressure variances the pressure dependence of permeability in laboratory and under *in situ* conditions varies from a few percent with respect to Figure 1 to more than 1 order *in situ* after ZIMMERMANN *et al.* (2005b). Some rock types show increasing and others decreasing permeability with fluid pressure under uniaxial strain conditions (Fig. 1). The anisotropy and heterogeneity of the rocks are obviously important to understand the variances of the properties.

In contrast to the *in situ* experiments, no threshold pressure indicating high permeability above and low permeability below was observed in the laboratory tests, even though fluid pressure variations were twice as large as for the field study. Some rock types show increasing and others decreasing permeability with fluid pressure under uniaxial strain conditions (Fig. 1). Therefore, we can conclude that we observed a lower percolation threshold pressure *in situ* than in laboratory at similar rocks, which showed only a hydraulic reaction to the fluid pressure variances. The critical values as discussed by GUEGUEN *et al.* (1997) are lower *in situ* and transferring laboratory findings needs appropriate relationships.

## *Conclusion*

Fluid pressure variations were performed on rock samples and during *in situ* tests of the reservoir rocks of Groß Schönebeck. Whereas a weak fluid pressure dependence of permeability was observed on rock samples, a stronger dependence was measured *in situ*. During the hydraulic fracturing treatments several sections of the reservoir could be stimulated, resulting in generation of multiple fractures (ZIMMERMANN *et al.*, 2005b) which connect different thicknesses of depth sections including the reservoir rock.

Within the reservoir several thresholds of fluid pressures for fracture opening were observed. Above the threshold high cumulative fracture conductivity of multiple fractures and below orders of magnitude lower fracture conductivity values were determined. The main parameter to obtain experimental data for critical threshold values is the injection rate because it affects the effective pressure. In the laboratory, no significant threshold was observed on core samples within comparable stress variations. Corresponding fluid pressures must be much higher.

The fluid pressure dependence of the transmissibility of the described reservoirs implies that self-propping of the fracture system after pressure release was not observed. However, the described mechanical behavior of the reservoir provides an opportunity for its utilization for geothermal fluid production from these reservoirs. The basic concept foresees a doublet of two boreholes, one for production and the other for injection. A fluid pressure close to the threshold realized in the injection well can be used as a controlling parameter for a reliable fluid production.

## *Acknowledgement*

The authors extend thanks to David Bruhn for critical reading and many suggestions to the manuscript and others from the geothermal project group (Ali Saadat, Silke Köhler, Helmuth Winter, and Heinz Holl a.o.) of the GFZ for many constructive discussions. We acknowledge the technical staff (e.g., Liane Liebeskind) for laboratory support and Jörg Schrötter and Matthias Poser for borehole measurements. The Groß Schönebeck project is a multidisciplinary joint venture of research institutes (GFZ Potsdam, BGR Hannover, GGA Hannover), Universities (TU Berlin, RU Bochum) and industry partners (GTN Neubrandenburg, BWG Neubrandenburg, MeSy Bochum). The annotations of two anonymous reviewers improved the paper. Many thanks go to the service companies (e.g. EEW and Schlumberger). Funds are provided by the German government from the ministries BMWI, BMBF, BMU, MWI Brandenburg and MWFK Brandenburg, recently under contract BMU ZIP 0327508 and BMU 0329951B.

## References


Baumgärtner, J., Jung, R., Hettkamp, T., and Teza, D. (2004), *The status of the hot dry rock scientific power plant at soultz-sous-forêts,* Z. Angew. Geol. *2*, 12–17.

Bourne, S.J. (2003), *Contrast of elastic properties between rock layers as a mechanism for the initiation and orientation of tensile failure under uniform remote compression,* J. Geophys. Res. B: Solid Earth. *108*(8), ETG 14-1–ETG 14-12.

Cinco Ley, H. and Samaniego, F. V. (1981), *Transient pressure analysis for fractured wells*, J. Petroleum Technology, 1749–1766.

Economides, M. J. and Nolte, K. G. (1989), *Reservoir Stimulation*, Schlumberger Educational Services, Houston, Texas.

Faulkner, D. R. and Rutter, E. H. (2003), *The effect of temperature, the nature of the pore fluid, and subyield differential stress on the permeability of phyllosilicate-rich fault gouge*, J. Geophys. Res. B: Solid Earth *108*(5): ETG 2-1–2-12.

Ghalambor Ali and Economides, M. J. (2002), *Formation damage abatement; a quarter-century perspective*, SPE J. *7*, 1, 4–13.

Gueguen, Y., David, C., and Gavrilenko, P. (1991), *Percolation networks and fluid transport in the crust*, Geophys. Res. Lett. *18*(5), 931–934.

Gueguen, Y., Chelidze, T., and Ravalec, M. L. E., (1997), *Microstructures, percolation thresholds, and rock properties*, Tectonophysics *279*, 23–35.

Gueguen, Y., Gavrilenko, P., and Ravalec, M. L. E. (1996), *Scales of rock permeability*, Survey in Geophys. *17*, 245–263.

Heiland, J., (2003), *Permeability of triaxially compressed sandstone: Influence of deformation and strain-rate on permeability,* Pure Appl. Geophys. *160*(5–6), 889–908.

Henninges, J., Zimmermann, G., Büttner, G., Schrötter, J., Erbas, K., and Huenges, E. (2005), *Wireline distributed temperature measurements and permanent installations behind casing*, Proc. World Geothermal Congress 2005, Antalya, Turkey, 24–29 April, 2005.

Huenges, E., Holl, H. G., Legarth, B., Zimmermann, G., Saadat, A., and Tischner, T. (2004), *Hydraulic stimulation of a sedimentary geothermal reservoir in the North German basin: Case study Groß Schönebeck*, Z. Angew. Geol. *2*, 24–27.

Legarth, B., Tischner, T., and Huenges, E. (2003), *Stimulation experiments in sedimentary, low-enthalpy reservoirs for geothermal power generation, Germany*, Geothermics *32*, Pergamon Press, 487–495.

Park, A. J, Tuncay, K., Laroche, M., Comer, J. B, and Ortoleva, P.J. (2001), *Comparing observed overpressuring and fracturing characteristics of siliciclastic and carbonate sediments and basin. RTM 3-D simulation predictions.* In american Association of Petroleum Geologists 2001 Annual Meeting: *Annual Meeting Expanded Abstracts*, Am. Assoc. Petroleum Geologists, 151. pp.

Trautwein, U. and Huenges, E. (2005), *Poroelastic behaviour of physical properties in Rotliegend Sandstones under uniaxial strain*, Internat. J. Rock Mechanics and Mining Sciences *42*, 7–8, 924–932.

Vinciguerra, S., Meredith, P. G., and Hazzard, J. (2004) *Experimental and modeling study of fluid pressure-driven fractures in Darley Dale sandstone,* Geophys. Res. Lett. *31*(9), L09609 1–4.

Yardley, B. W. D. (2001), *Rock controls on fluid regimes in the crust,* Yardley Bruce W. D. Earth system processes; programmes with abstracts, 53 pp. Geolog. Soc. Am. Geolog. Soc. London, International.

Zimmermann, G., Reinicke, A., Holl, H. G., Legarth, B., Saadat, A., and Huenges, E. (2005a), *Well test analysis after massive waterfrac treatments in a sedimentary geothermal reservoir,* Proc. World Geothermal Congress 2005, Antalya, Turkey, 24–29 April, 2005.

ZIMMERMANN, G., BURKHARDT, H., and ENGELHARD, L. (2003), *Scale dependence of hydraulic parameters in the crystalline rock of the KTB*, Pure Appl. Geophys. *160*, 1067–1085.

ZIMMERMANN, G., BURKHARDT, H., and ENGELHARD, L. (2005b), *Scale dependence of hydraulic and structural parameters in fractured rock, from borehole data (KTB and HSDP)*. In *Petrophysical Properties of Crystalline Rocks* (Harvey, P.K., Brewer, T.S., Pezard, P.A., and Petrov, V.A., eds), Geolog. Soc. London, Special Publications *240*, 37–45.

ZOBACK, M.D., BARTON, C.A., BRUDY, M., CASTILLO, D.A., FINKBEINER, T., GROLLIMUND. B.R., MOOS, D.B., PESKA, P., WARD, C.D., and WIPRUT, D.J. (2003), *Determination of stress orientation and magnitude in deep wells,* Internat. J. Rock Mechanics and Mining Sciences. *40*(7–8), 1049–1076.

(Received March 24, 2005, revised February 20, 2006, accepted March 5, 2006)
Published Online First: September 14, 2006

Pure appl. geophys. 163 (2006) 2153–2170
0033–4553/06/102153–18
DOI 10.1007/s00024-006-0123-1



Pure and Applied Geophysics

# Characterisation of Tensile Damage in Rock Samples Induced by Different Stress Paths

PATRICK GANNE and ANDRÉ VERVOORT

*Abstract*—Cracks, induced by different stress states, are studied systematically on thin slices of crinoidal limestone samples. Samples damaged by tension are compared to samples in which first damage is induced by compressive stresses, followed by tensile stresses. To quantify the effect of the compressive stresses, samples only damaged by compression are also studied. The tensile stresses cause the occurrence of an intergranular crack. When the sample is first damaged by compressive stresses, former induced damage influences the growth of this crack. It results in an intergranular crack with a zigzag profile. The average orientation of the intergranular crack is, however, the same as in the samples damaged by tension only.

**Key words:** Micro-damage, fractures, stress path, tensile stresses, crinoidal limestone.

## *1. Introduction*

The strength of material is affected by its micro-scale quality such as the presence of cracks and fissures. An increase of the overall stress level also causes higher local stresses. If the local stress exceeds the local strength, cracks occur, grow and interact, decreasing the local coherence (KRANZ, 1983). When the orientation or the magnitude of the principal stresses changes, new damage may occur. The amount of new damage depends, apart from the stress level and the stress orientation, on the damage that has previously been induced. In other words, the new damage is influenced by the past stress path.

In order to predict the stability of the rock mass around excavations, rock samples are tested as input for the design based on some general theories and modelling. During this process, the effect of the stress path on micro-damage and, hence, on the strength and on the rock behaviour is normally not taken into account. As suggested by e.g. WANG *et al.* (1993), DER-HER LEE *et al.* (1999), HOLT *et al.* (2000) and KAISER *et al.* (2001) the stress path has an effect on the stability of the

Research Unit Mining, K.U.Leuven, Kasteelpark Arenberg 40, B-3001Leuven, Belgium.
E-mail: Patrick.ganne@bwk.kuleuven.be

rock mass. As the principal stress level and orientation change significantly around mining and other geotechnical activities (EBERHARDT, 2001), fundamental research of the influence of the stress path is important.

The paper describes and discusses the influence of the stress path on micro-damage. Two extreme situations are considered. First, the tensile damage on a fresh sample is studied. Second, a sample that has undergone compressive damage is loaded in a tensile mode. Both conditions of tensile damage are compared and the effect of the prior compressive damage is quantified.

Most of the study is based on a systematic observation of thin slices. To create appropriate tensile and compressive damage a specific sample geometry is used, as described in the following paragraph.

## 2. Methodology

The rock material used in the tests is Belgian crinoidal limestone. It dates back to the Dinantian period of the Carboniferious area, and more particularly to the Tournesian period (368-359 million years ago, DESCHUYTENER, 1984). This material is mainly composed of calcite, in various forms. Besides crinoids, the limestone also contains other bioclasts. Another major constituent of the crinoidal limestone is micrite. This very fine-grained microcrystalline calcite makes up the matrix of the limestone. Stylolites are frequently encountered. Test samples, containing outspoken stylolite bands, are rejected for testing.

The crinoidal limestone blocks are drilled, sawed and rectified into rectangular slabs of 31 mm × 60 mm × 140 mm. At one side, half a cylinder is drilled with a diameter of 58 mm (Fig. 1). The loading direction determines the stress redistribution. If the external load is applied on the short sides of the rectangular samples, compressive stresses are present adjacent to the borehole. Fig. 2a shows the calculations (Flac® 4.0, Itasca Consulting Group, Inc.) of the stress redistribution on the symmetrical plane when a loading force of about 80 kN (about 75% of the failure load) is applied on the short sides of a sample. This is further called the compression configuration. If the external load is applied on the long sides, both the radial and tangential stresses are negative (tensile stresses) next to the borehole. Fig. 2b shows the calculations of the stress redistribution on the symmetrical plane when the load on the long sides of a sample rises until plastic deformation occurs (at an external load of about 40 kN). It has been checked that the zone of tensile stresses extends at the borehole from position angle −45° until +45° (the definition of 'position angle' is given further on in this paragraph). This is further called the tension configuration.

The Dartec® loading machine is used in the displacement-controlled mode. During the tests, 1 mm thick Teflon is placed between the machine compression platens and the test sample, in order to minimize the frictional stresses. Acoustic

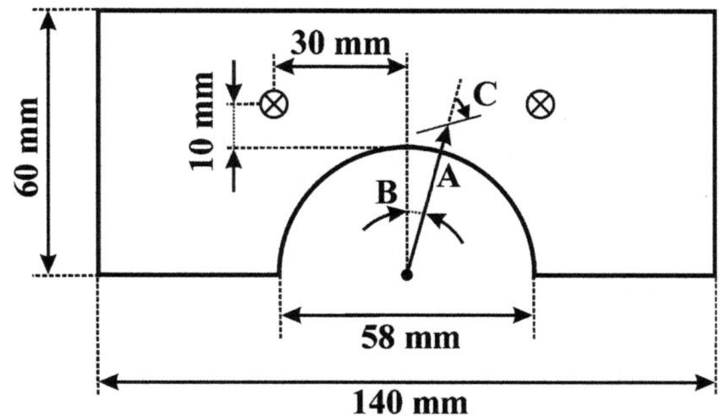


Figure 1
Geometry of the tested samples, location of the acoustic emission (AE) sensors (⊗) and parameters describing a crack (radial distance A, position angle B and orientation C, i.e., the orientation relative to the radial direction).

emissions (AE) are continuously recorded during each loading cycle using the AMS3 system of Vallen-System GmbH. Two wide-band AE sensors of type B1025 (Digital Wave Corp.) with a frequency range from 50 kHz to 2 MHz are attached to the test samples, symmetrically around half the borehole (Fig. 1). The distance between the AE sensors in the compression configuration and in the tension configuration is about 60 and 40 mm, respectively. This set-up allows a linear localization of AE sources. The velocity of the longitudinal elastic wave used for the location calculations is taken as 5700 m/s. LAVROV *et al.* (2002) suggest that the anisotropy of the velocity in the Belgian crinoidal limestone does not exceed 5%. Low anisotropy of the velocity is one of the prerequisites for the location reliability. Thus, it is possible to separate the AE events occurring in the central zone from AE events outside this zone of interest.

In the compression configuration (Fig. 3), the cumulative number of AE events clearly shows the moment of first cracks formation, the zone of the systematic damage and of crack interaction, as defined by CAI *et al.* (2004). The change of the steepness of this curve defines the different zones. DIEDERICHS *et al.* (2004) suggest that only the zones of systematic damage and of the crack interaction have physical interpretations: First individual microcracks occur (systematic damage zone), second cracks interact and a macro-fracture occurs (crack interaction zone). In the test samples, the systematic damage zone starts at 55% ± 5% of the failure load, while the crack interaction zone starts at 85% ± 5%. Figure 3 gives the cumulative number of events versus applied load in a compression configuration until failure. For the experiments presented further in this paper, the loading is stopped in the zone of the systematic damage (grey vertical line in Fig. 3).

In the tension configuration, a cotton wool, saturated with water, is put inside the borehole. The loading is stopped when a crack of several centimetres is visualised by the capillarity water suction into the crack and when numerous AE events are

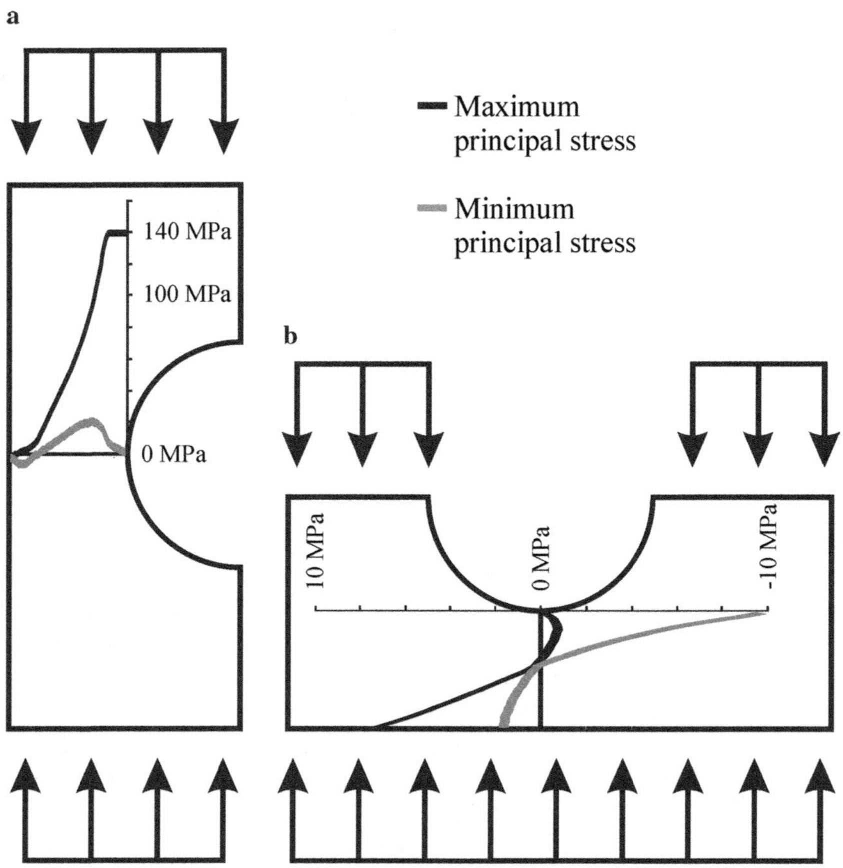


Figure 2
Stress redistribution on the symmetrical plane for the two loading directions (a: 'compression' configuration with a loading force of 80 kN, i.e., about 75% of the failure load; b: 'tension' configuration at the beginning of plastic deformation; positive stresses are compressive).

registered. This crack, visualised by capillarity water, would not be visible by the naked eye without water suction. Van de Steen (2001) used this technique and showed that there was no effect of the water on the crack formation.

To quantify the damage in tension only, a load is applied on the long sides (tension configuration, further called T-sample). To quantify the effect of damage in compression, followed by damage in tension, the load is first applied as in the compression configuration before the sample is rotated by 90° (tension configuration). This loading procedure is further called CT (two samples tested: CTa and CTb). Between both loading cycles, the samples are not loaded during several days. To quantify the damage by compression only, a load is applied as in the compression configuration (further called C-sample).

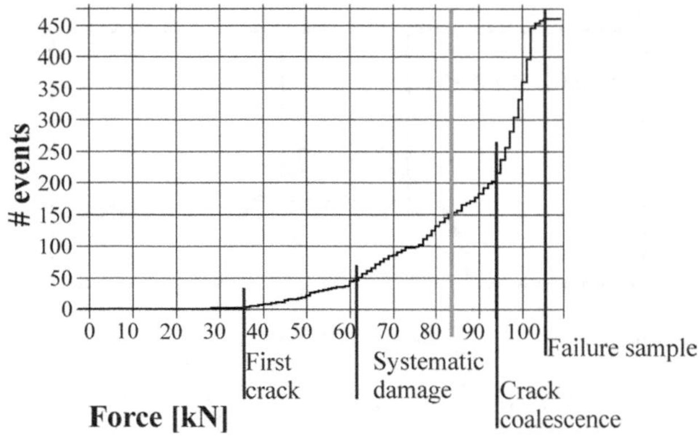


Figure 3
Number of cumulative AE-events during a loading test with a 'compression' configuration. For the experiments further discussed in this paper, the loading is stopped at the grey line.

Prior to preparing thin slices, the samples are impregnated under vacuum with a fluorescent resin. Thin slices with a thickness of 30 μm are prepared. Figure 4 shows the position of the thin slice in each sample. The thin slice is taken in the middle of the thickness. This is the most likely damaged region. A region of 20 mm tangentially and 6 mm wide (radial direction) is digitally photographed with a pixel resolution of 0.5 μm. By composing semi-automatically one mosaic photograph, the pixel accuracy decreases to 1%. The cracks are semi-automatically recorded in the final photograph. As the cracks are not always easy to observe on the digital photographs, manual decisions are taken after viewing through the microscope. Some cracks are only visible through transmitted light in the microscope (x5), while others are only visible under a reflection fluorescence microscope (x5). For each visible crack (Fig. 1), the positions of the central and the end points (defined by the radial distance, A, and the angle, B, of the position), the orientation (i.e., the absolute value of the angle between the orientation of the crack and the radial direction, C) and the length are recorded. As the direction of an intergranular crack frequently changes, the intergranular cracks are subdivided into connected multiple smaller (length > 15 μm) straight cracks. Because of the symmetry of the sample configuration, the external loading and the stress redistribution, only the absolute value of the orientation is considered. The features of cracks crossing micrite or crystals could only be determined by manual inspection under the microscope. Matlab® performs the calculation for these parameters, as well as the statistical analyses. The definitions, as well as the calculation methodologies are in correspondence to Russ (2002). An interesting feature in the statistical analyses is for example the total length of cracks within a specific orientation range: the length of all cracks orientated

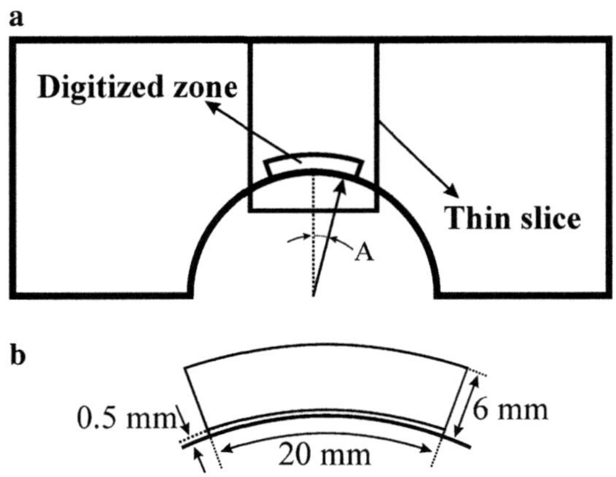


Figure 4
a: Position of the thin section (rectangle) and the digitized region and the indication of the position angle (A); b: Detailed view of the digitized region and the indication of the 0.5 mm band which is not taken into account.

within this specific orientation range is summed. This information is summarized in a graph of the total length of all cracks within which crack orientation ranges as a function of these orientation ranges.

The various C-, T-, CTa- and CTb-samples are analysed in this way, as well as an unloaded (U-) sample. To study the possible effects of sawing, rectifying, drilling, thin slice preparation, etc., these actions are also performed on the U-sample, however no loading is applied.

As is shown further, numerous cracks are activated cleavage planes of individual crystals. Most of these cracks are intragranular cracks. KRANZ (1983) gives a detailed explanation and definition of the used terminology. Figure 5 shows examples of different encountered features. Figure 5a shows a crystal with several activated cleavage planes. Also the activated twins are encountered as cracks. Most of these cracks do not show the presence of impregnated fluorescent resin. Other types of observed cracks are impregnated cracks. The resin penetrates the cracks which are more open. These impregnated cracks are visible under the reflection fluorescence microscope. Some damage is induced during the preparation of the thin slices (Fig. 5b), but is clearly recognized as such. Such damage is not taken into account in the further analyses. Some anomalies could incorrectly be interpreted as cracks. For example, some bioclasts are very similar to cracks (Fig. 5c). Furthermore some grain boundaries, and specifically grain boundaries of dolomitised crystals, are difficult to distinguish from real cracks (Fig. 5d). Only the features that are not only situated at grain boundaries and are not bioclasts are further considered for the analyses. In these cases, a direct observation through the microscope (x5) allows a correct interpretation.

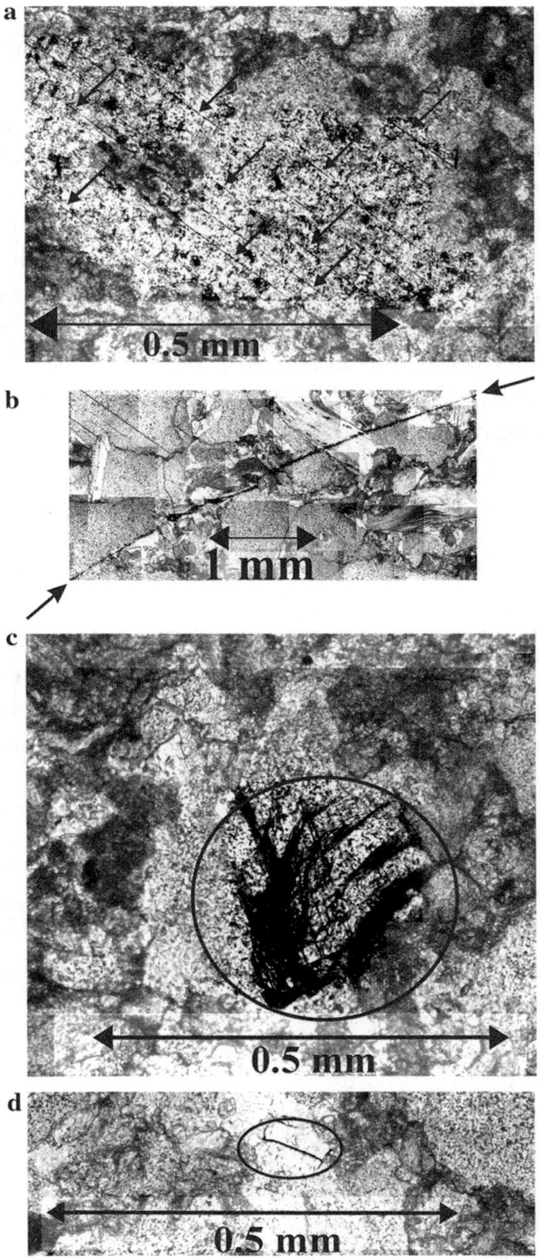


Figure 5
Examples of a: activated cleavage planes in a crystal; b: a crack induced by the preparation of the thin slice (between both arrows); c: a bioclast; d: grain boundary of dolomitised crystals.

It should be stressed that due to parallel lying activated cleavage planes, the total length of cracks appears to be high. For example the one crystal in Fig. 5a with a diameter of about 0.6 mm, has several activated cleavage planes. These increase the total crack length by about 3.5 mm.

In some regions of the thin slices, the visibility is poor. It depends on the purity of the crystals, the darkness of the micrite, the presence of organic material, etc. In some cases, cracks cannot be observed and hence, they are not taken into account. Due to the fluorescent resin, the impregnated cracks do not deal with this problem. The percentage of the useful surface varies slightly for the different samples, between 85 and 91% (Table 1). The error margin on this percentage for each sample is estimated at about 4%. In other words, the variation of the percentage of useful surface over the different samples barely exceeds the error margin. Hence, no correction is further applied for the different percentages.

## *3. Results and Interpretations*

In all samples, including the U-sample, numerous cracks occur within a 0.5 mm band around the borehole. The total length of all cracks within this band varies between 26 and 58 mm. In this region, the dominant orientation of the cracks is 80° ± 10° (in other words with a tangential orientation). These cracks, within this band of 0.5 mm, are interpreted as a result of the drilling of the borehole (Fig. 6). VAN DE

Table 1

*Statistical overview of the encountered non-impregnated cracks and the useful surface percentage in the U-, C-, T- and CT-samples in a region of about 20 mm along the borehole and from 0.5 to 6 mm away from the borehole in the radial direction*

| Useful surface percentage | U | C | T | CTa | CTb |
|---|---|---|---|---|---|
| | 86 ± 4 % | 85 ± 4 % | 88 ± 4 % | 89 ± 4 % | 91 ± 4 % |
| # non-impregnated cracks | 706 | 1275 | 2004 | 1325 | 1693 |
| Min. length of non-impregnated cracks | 0.013 mm | 0.012 mm | 0.013 mm | 0.013 mm | 0.013 mm |
| Max. length of non-impregnated cracks | 0.389 mm | 1.421 mm | 0.479 mm | 0.396 mm | 0.642 mm |
| Arithmetic mean of the length of non-impregnated cracks | 0.045 mm | 0.068 mm | 0.041 mm | 0.065 mm | 0.061 mm |
| % non-impregnated cracks > 0.1 mm | 5 % | 14 % | 3 % | 15 % | 11 % |
| Length % of non-impregnated cracks > 0.1 mm | 18 % | 39 % | 13 % | 34 % | 33 % |
| # impregnated cracks* | 20 + 0 | 25 + 0 | 39 + 3 | 62 + 1 | 94 + 1 |
| Min. length of impregnated cracks | 0.040 mm | 0.016 mm | 0.020 mm | 0.015 mm | 0.018 mm |
| Max. length of impregnated cracks | 0.933 mm | 0.922 mm | > 6 mm | > 6 mm | > 6 mm |
| % impregnated cracks > 0.1 mm | 85 % | 92 % | 83 % | 87 % | 72 % |

* the intergranular cracks are shown after the plus sign.

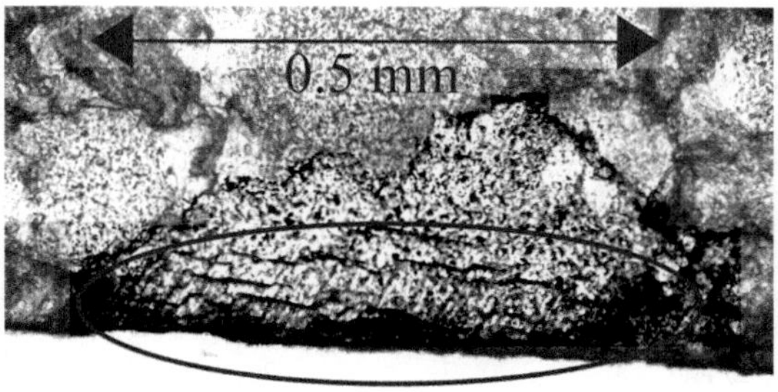


Figure 6
Example of cracks at the border of the borehole.

Figure 7
Example of an intergranular crack.

STEEN (2001) observes similar features while drilling under laboratory conditions in crinoidal limestone. Although the damage in this band affects the occurrence and further growth of cracks, and although not all cracks within this region are caused by the drilling, it is decided not to consider this 0.5 mm band in the further analyses (Fig. 4).

In the digitized region, a large number of non-impregnated cracks are visible under microscope (varying from 706 to 2004 cracks for all samples studied; Table 1).

Table 2

*Overview of the observed total length of all cracks, impregnated and non-impregnated cracks in the U-, C-, T- and CT-samples in a region of about 20 mm along the borehole and from 0.5 to 6 mm away from the borehole in the radial direction*

| | U<br>mm | C<br>mm | T<br>mm | CTa<br>mm | CTb<br>mm |
|---|---|---|---|---|---|
| Σ length of all cracks | 37 | 106 | 109 | 115 | 134 |
| Σ length of impregnated cracks | 5 | 20 | 33 | 32 | 40 |
| Σ length of non-impregnated cracks | 32 | 83 | 76 | 83 | 94 |

The arithmetic mean of the length of the non-impregnated cracks varies between 0.041 mm and 0.068 mm; the minimum and the maximum observed length of these non-impregnated cracks are respectively 0.012 mm and 1.421 mm.

A smaller number of impregnated cracks are observed (between 20 and 95). The shortest observed impregnated crack is 0.015 mm. Some of these impregnated cracks are intergranular cracks with a length of several centimetres (Fig. 7). Only the parts of the intergranular cracks lying within the digitized region are taken into account (see Fig. 4b). By this procedure, the total length of all impregnated cracks within the digitized region varies between 5 and 40 mm in all samples studied (Table 2, Fig. 8). The number of observed cracks (impregnated and non-impregnated) justifies thus a statistical analysis.

### *3.1 U- and C-samples*

Although the U-sample has not undergone a loading cycle, cracks appear under the transmitted light microscope as well as under the reflection fluorescence microscope (in total 706 non-impregnated and 20 impregnated cracks). The total length of all impregnated cracks is 5 mm and of all non-impregnated cracks is

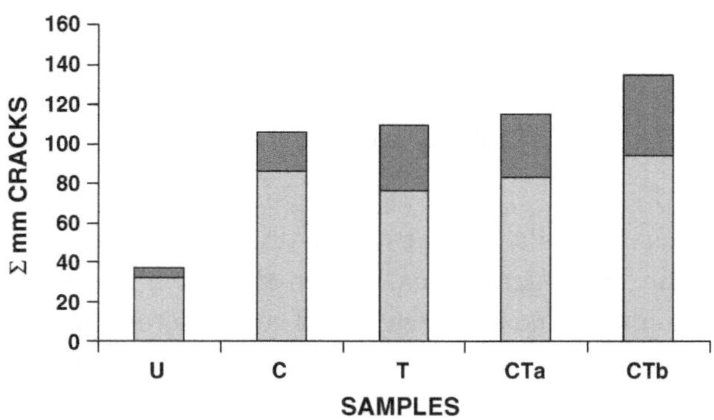


Figure 8
The total length of impregnated cracks (dark grey) and of non-impregnated cracks (light grey).

32 mm. Some of these are natural cracks (Fig. 9), while others are due to the preparation of the test sample and the thin slice (Fig. 5b, Fig. 6).

The arithmetic mean of the length of the non-impregnated cracks is 0.045 mm and only 5 % of the non-impregnated cracks are longer than 0.1 mm (Table 1). Seventeen of the 20 impregnated cracks are longer than 0.1 mm (maximum length of 0.933 mm). No specific orientation is dominantly present among the cracks.

As mentioned above, the C-sample is not loaded until failure, but the loading is stopped within the zone of systematic damage (Fig. 3). More non-impregnated and impregnated cracks are observed in the C-sample (1275 and 25, respectively) than in the U-sample (706 and 20, respectively). In the C-sample, a total of 20 mm impregnated and 83 mm non-impregnated cracks are observed. All these cracks are intragranular. Fig. 10a shows the cracks in the C-sample between 0° and 8° position angle. Typically, these cracks are activated cleavage planes of individual crystals (Fig. 5a).

These cracks are scattered over the entire digitized region. Most of the impregnated cracks occur within crystals where the orientation of the natural cleavage planes deviate about 60° from the maximum principal stress, that being the tangential stress. The total length of all cracks within crack orientation ranges (Fig. 11a) is for all ranges larger than for an unloaded sample (U) except perhaps orientations larger than 75°. As mentioned above, the absolute value of the crack orientations are taken into account. Cracks with an orientation of 60° ± 10° are dominant at the position angle −4° to 4°. If the calculated stress situation is considered, the general theory of Mohr-Coulomb predicts macro-fractures with an orientation of about 60° from the radial direction.

The number of cracks longer than 0.1 mm increases from 5% (in the U-sample) to 14% (in the C-sample, Table 1). By also considering the increase of the number of cracks in comparison to the U-sample, one can conclude that, due to the loading, an

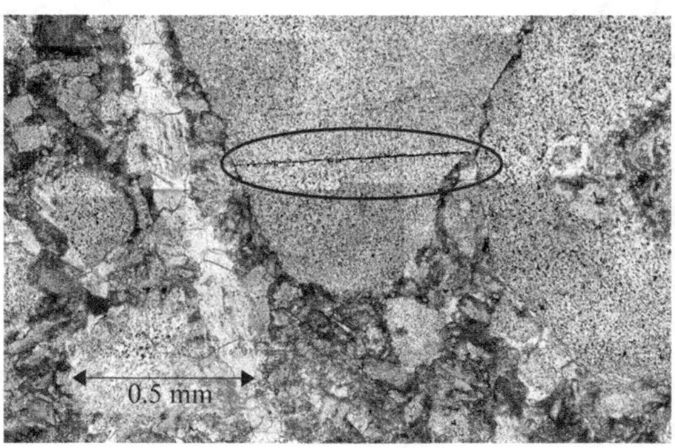


Figure 9
Example of a pre-existing crack in the unloaded sample (U).

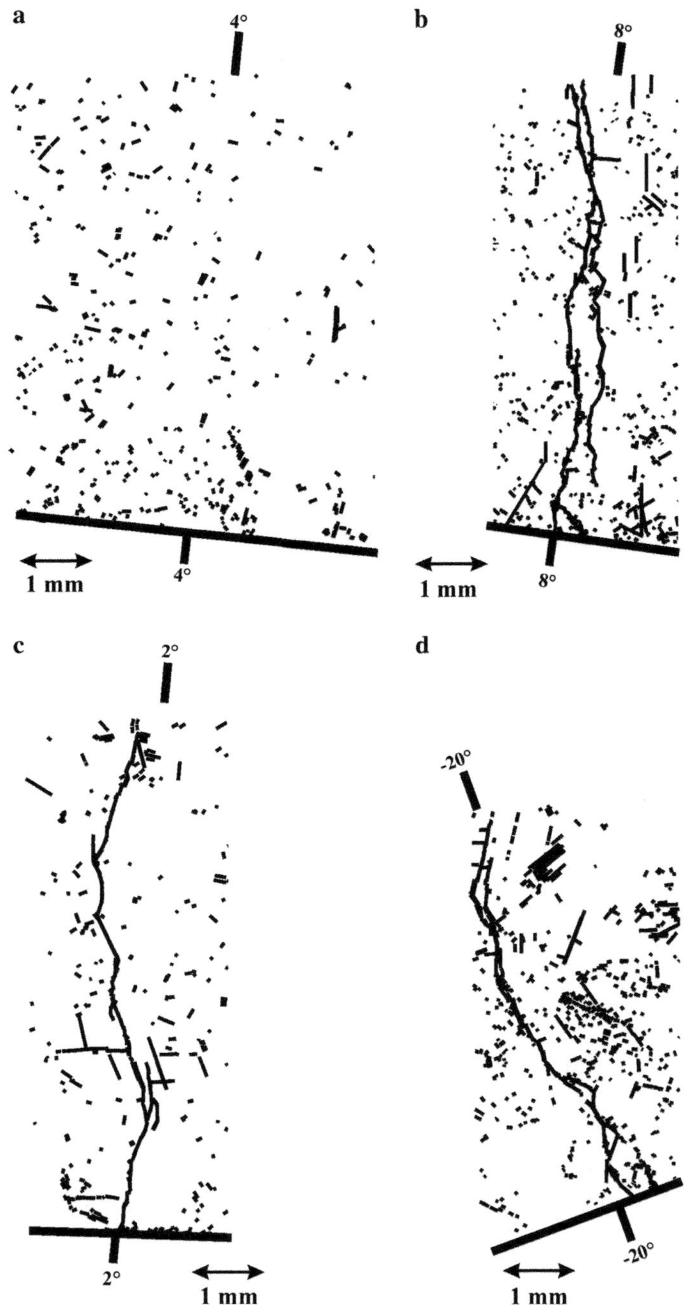


Figure 10
View of cracks and intergranular cracks in (a) the C-sample between position angle 0° and 8°, (b) T-sample between position angle 6° and 11°, (c) CTa-sample between position angle 0° and 4° and (d) CTb-sample between position angle −24° and −18°.

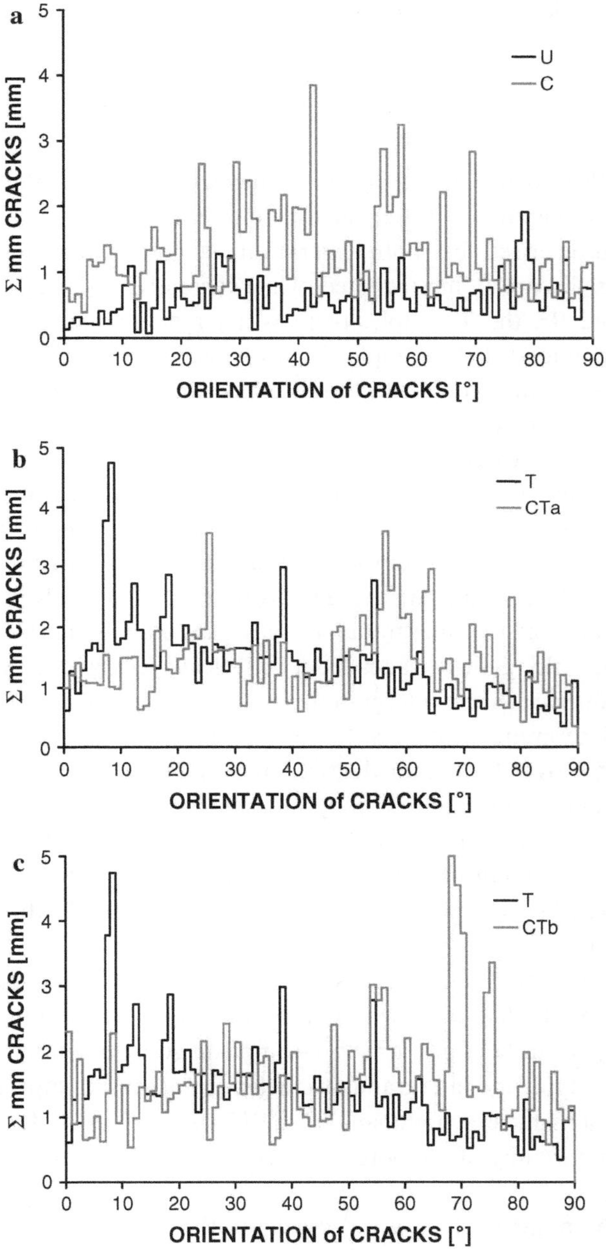


Figure 11
The total length of all cracks (impregnated, as well as non-impregnated) within crack orientation ranges as a function of these orientation ranges of (a) the U- and C-samples, (b) the T- and CTa-samples, (c) the T- and CTb-samples. As mentioned in the text, only the absolute values of the orientations are considered.

abundance of cracks occur as well as numerous cracks that have grown within the digitized zone.

### *3.2 T-sample*

The T-sample is loaded in the tension configuration until an intergranular crack is visualised by the capillarity water suction into this crack. In this T-sample, the damage is characterised by both intergranular and intragranular cracks. The intergranular cracks are all impregnated, while generally the intragranular cracks are non-impregnated. In the T-sample, there is a larger amount of non-impregnated cracks as in the unloaded (U) sample (2004 versus 706, see Table 1). In comparison to the U-sample (5 mm impregnated and 32 mm non-impregnated cracks), the total length of the impregnated (33 mm) and of non-impregnated cracks (76 mm) in the digitized region is increased (Fig. 8).

Intergranular cracks go along grain boundaries as well as along cleavage planes of crystals. VAN DE STEEN *et al.* (2002) observed in Brazilian tested samples of crinoidal limestone that the proportion of crinoids along a fracture path, caused by tensile stresses, is larger than the average proportion of crinoids in the tested rock. This would mean that the intergranular cracks, induced by tensile stresses, have a preference to go along the weaker bounded cleavage planes. Typically at grain boundaries or in micrite, the intergranular cracks branch off.

The long intergranular cracks (3) are radially oriented. Two intergranular cracks are visible over the 6 mm digitized region. At the position angle of about 8°, these two intergranular cracks along most of their length parallel to each other at a distance of about 0.2 mm (Fig. 10b). At the position angle −12° and close to the borehole, an intergranular crack of about 1.2 mm is observed. The intergranular cracks represent about 85 % of the total observed length of impregnated cracks.

The observed intergranular cracks have a radial (=0°) orientation (Fig. 10b). The total length of all cracks within the crack orientation range of 6° to 10° (Fig. 11b) is significantly larger for the T-sample in comparison to the U-sample. In the 0.4 mm vicinity of the intergranular cracks, the observed impregnated and non-impregnated cracks have a dominant 0° ± 10° orientation. If the principal stress orientation in the vicinity of the borehole is considered, radial tensile macro-fracture seems logical.

Although the number of observed cracks in the T-sample is higher than in the U-sample, the arithmetic mean of the length of non-impregnated cracks and the percentage of non-impregnated cracks longer than 0.1 mm are comparable (0.041 mm versus 0.045 mm and 3% versus 5%, see Table 1). It has been verified that 62% of the cracks longer than 0.1 mm are situated around (±1 mm) the intergranular cracks.

### 3.3 *CTa- and CTb-samples*

As already mentioned, the CT-samples are first loaded in the compression configuration until the systematic damage zone is reached. Hereafter, the CT-samples are loaded in the tension configuration until an intergranular crack is visualised by capillarity water suction. Both intergranular and intragranular cracks are observed. Again, the intergranular cracks are all impregnated, while most of the intragranular cracks are non-impregnated. In total, 83 mm to 94 mm (U-sample: 32 mm) of non-impregnated cracks and 32 mm to 40 mm (U-sample: 5 mm) of impregnated cracks are observed.

The intergranular cracks go along grain boundaries as well as along cleavage planes of crystals. It seems that in comparison to the T-sample, the intergranular cracks in the CT-samples more frequently follow the weaker bounded cleavage planes. These cracks are visible over the 6 mm digitized region. In both samples, only one intergranular crack is observed: in the CTa-sample at an average position angle of 2° and in the CTb-sample at −20° (Figs. 10c and d). In the digitized regions, the intergranular cracks make up about 65 and 70 length % of the total impregnated cracks.

Although the intergranular cracks have a zigzag profile (Figs. 10c and d), the average orientation of the intergranular cracks is radial. The total length of all cracks within crack orientation ranges shows an increase in the scope of 58° to 65° (CTa, Fig. 11b) and in the 56° to 75° extent (CTb, Fig. 11c). A minor increase is visual around 25°. It has been verified that the cracks with an orientation of 20° ± 10° are dominant at the position angle of the intergranular cracks. The cracks with an orientation of 60° ± 10° are scattered over the entire digitized region. In the CTa-sample, these cracks are dominant both at −8° to −2° and at 2° to 8° position angles. In the CTb-sample, these cracks are dominant at the −4° to 4° position angle, and at the −17° to −13° position angle (as mentioned above, the intergranular cracks are at 0° and at −20°, respectively).

In comparison to the U-sample, the percentage of non-impregnated cracks larger than 0.1 mm is increased from 5% to respectively 15% and 11% (Table 1). Accordingly, the arithmetic mean of the length of the non-impregnated cracks increases from 0.045 mm to 0.065 mm and 0.061 mm.

## 4. *Discussion: Influence of the Compressive Damage on Fracturing in Tension*

The damage caused by the loading cycle with the tension configuration in the T-sample and the CT-samples should be interpreted allowing for earlier damage. In the T-sample, the pre-loading damage, which could be visualised on thin slices (U-sample) amounts to a total length of about 37 mm cracks within the entire digitized region (122 mm$^2$). In the CT-samples, the pre-loading damage

(C-sample) amounts to about 106 mm cracks within the entire digitized region. This total length is composed of about 1275 non-impregnated and 25 impregnated cracks.

In the T-sample, the loading causes a large amount of non-impregnated cracks (a total length of about 44 mm) as well as a total of about 28 mm impregnated cracks within the digitized region. Although the number of non-impregnated cracks increases (from 706 (U-sample) to 2004), the arithmetic mean of non-impregnated cracks and the percentage of non-impregnated cracks longer than 0.1 mm are comparable (0.045 mm versus 0.041 mm; 5% versus 3%, respectively). Characteristic of the additional impregnated cracks is the occurrence of 3 intergranular cracks, of which 2 intergranular cracks are several centimetres long. These intergranular cracks are radially oriented. The total length of all cracks within the crack orientation range of 6° to 10° (Fig. 11) is larger for the T-sample in comparison to the U-sample. Within a band of 0.4 m around the intergranular cracks, the observed impregnated and non-impregnated cracks have a dominant 0° ± 10° orientation.

In the CT-samples, the loading in the tension configuration causes (in comparison to the former C-damage) about 0 to 11 mm additional non-impregnated cracks and 12 to 20 mm impregnated cracks. Although the number of non-impregnated cracks increases (from 1275 (C-sample) to 1325 and 1693), the arithmetic mean of non-impregnated cracks and the percentage of non-impregnated cracks longer than 0.1 mm are comparable (0.068 mm versus 0.065 mm and 0.061 mm; 14% versus 15% and 11%, respectively). Characteristic for the additional impregnated cracks is the occurrence of intergranular cracks which are several centimetres long. Although the intergranular cracks have a zigzag profile, they are on average radially oriented. The total length of all cracks within a crack orientation range (Fig. 11b) increases slightly around 25° orientations (CTa). This increase is hardly noticeable for the CTb-sample. In a zone of about 0.4 mm around the intergranular cracks, the observed impregnated and non-impregnated cracks have a dominant 20° ± 10° orientation. These orientations deviate more from the radial orientation in the CT-samples than in the T-sample.

In the CT-samples, the intergranular cracks are clearly influenced by the former induced cracks, which are generally not radially oriented. The intergranular cracks frequently find former activated cleavage planes as easy ways to grow. By this, the intergranular cracks grow more along already activated cleavage planes than the intergranular cracks in the T-sample do. A comparison of Figs. 10c and d (intergranular crack in CT-samples) with Fig. 10b (intergranular crack in T-sample) indicates that the intergranular cracks in the CT-samples have more frequently straight parts (between 0.2 mm and 0.6 mm long). These parts are cleavage planes of individual crystals. The frequent use of activated cleavage planes, which are mostly not radially oriented, causes the zigzag profile of the intergranular cracks in the CT-samples.

Because the intergranular crack uses existing cracks more in the CT-samples than in the T-sample, the less observed additional damage seems logical.

## *5. Conclusion*

By a systematic study based on thin slices, the effect of prior compressive damage on the occurrence of a tensile macro-fracture is visualised. The tensile fracture follows more cleavage planes than in a sample without prior loading. The cleavage planes that affect the fracture path are activated in a compressive stress field and most of them are not oriented radially. With and without prior loading, the average orientation of the tensile macro-fracture is radial. In the samples with prior compressive damage, the tensile macro-fracture has a clear zigzag profile around the average radial direction. Around these macro-fractures, short parallel fractures are observed.

It is obvious that these observations are clearly affected by the typical structure of the crinoidal limestone with its typical cleavage planes. Hence, the conclusions cannot be extrapolated directly to other rock types. However, the methodology applied is suitable for use with similar investigations of other rock types and/or other loading paths.

## References

CAI, M., KAISER, P.K., TASAKA, Y., MAEJIMA, T., MORIOKA, H., and MINAMI, M. (2004), *Generalised crack initiation and crack damage stress thresholds of brittle rock masses near underground excavations*, Int. J. Rock Mech. Min. Sci. *41*, 833–847.

DER-HER LEE, JUANG, C.H., JIN-WEN CHEN, HUNG-MING LIN., and WEN-HWANG SHIEH (1999), *Stress paths and mechanical behavior of a sandstone in hollow cylinder tests*, Int. J. Rock Mech. Min. Sci. *36*, 857–870.

DESCHUYTENER, M.L., PETIT GRANIT, Technical note (in French) (CSTC, Brussels, 1984).

DIEDERICHS, M.S., KAISER, P.K., and EBERHARDT, E. (2004), *Damage initiation and propagation in hard rock during tunnelling and the influence of near-face stress rotation*, Int. J. Rock Mech. Min. Sci. *41*, 785–812.

EBERHARDT, E. (2001), *Numerical modelling of three-dimension stress rotation ahead of an advancing tunnel face*, Int. J. Rock Mech. Min. Sci. *38*, 499–518.

HOLT, R.M., BRIGNOLI, M., and KENTER, C.J. (2000), *Core quality: Quantification of coring induced rock alteration*, Int. J. Rock. Mech. Min. Sci. *37*, 889–907.

KAISER, P.K., YAZICI, S., and MALONEY, S. (2001), *Mining-induced stress change and consequences of stress path on excavation stability – A case study*, Int. J. Rock Mech. Min. Sci. *38*, 167–180.

Kranz, R.L. (1983), Microcracks in rocks: A review, Tectonophysics *100*, 449–480.

LAVROV, A., VERVOORT, A., WEVERS, M., and NAPIER, J.A.L. (2002), *Experimental and numerical study of the Kaiser effect in cyclic Brazilian tests with disk rotation*, Int. J. Rock Mech. Min. Sci. *39*, 287–302.

RUSS, J.C., *The Image Processing Handbook,* Fourth Edition (CRC Press, London 2002).

VAN DE STEEN, B., VERVOORT, A., and SAHIN, K. (2002), *Influence of internal structure of crinoidal limestone on fracture paths*, Eng. Geol. *67*, 109–125.

Van de Steen, B, *Effect of heterogeneities and defects on the fracture pattern in brittle rock* (Ph.D. Thesis, K.U. Leuven, Leuven 2001).

Wang, Y., Kessler, N., and Santarelli, F.J. (1993), *Stress-induced anisotropy and loading path effects on borehole responses*, Int. J. Rock Mech. Min. Sci. and Geomech. Abst. *30*, 1367–1370.

(Received March 2, 2005, revised November 15, 2005, accepted November 16, 2005)
Published Online First: September 14, 2006

Pure appl. geophys. 163 (2006) 2171–2204
0033–4553/06/102171–34
DOI 10.1007/s00024-006-0117-z



Pure and Applied Geophysics

# Monitoring Fracture Propagation in a Soft Rock (Neapolitan Tuff) Using Acoustic Emissions and Digital Images

Stephen A. Hall,[1] Fiorenza de Sanctis,[1,2] and Gioacchino Viggiani[1]

*Abstract*—Sudden and unexpected collapses of underground cavities below the city of Naples (Italy) represent a serious safety hazard. The collapses occur due to the detachment of large blocks from the cavity roofs, walls and pillars, often a long time after the original quarry excavation has been completed. It is recognised that existing discontinuities, e.g., fractures, play an important role in the failure process by inducing local stress-concentrations and reducing the overall material strength. The larger fractures, which ultimately lead to collapse occur through interaction, propagation and coalescence of these discontinuities. This paper presents recent results of experiments carried out on natural, dry specimens of Neapolitan fine-grained tuff to investigate the mechanisms involved in sample failure. A better understanding of fracture development and rock bridge behaviour is gained through a combination of AE and photographic monitoring in an experimental program considering samples with artificial pre-existing heterogeneities, which simulate the *in situ* discontinuities. For a range of rock bridge geometries the mechanisms and timing of different stages of the failure process are identified and characterised. The results show that, in general, a classical description of failure, for samples without artificial flaws or with only a single flaw, is followed: (1) crack closure; (2) linear stress-strain response and crack initiation with stable crack growth; (3) crack damage and unstable crack growth leading to failure. For samples with two artificial pre-existing flaws the third phase is split into two parts and failure of the sample occurs only after both the unstable propagation of external wing cracks and coalescence of the internal cracks in the bridge. In terms of the timing and duration of each phase, it is seen that phases 1 and 2 have little dependence on the flaw configuration but phase 3 seems to depend directly on this. In particular the angle in rock bridge between the inner tips of the pre-existing flaws, $\beta$, plays a key role: phase 3 is shorter for $\beta = 120°$ than for $\beta = 105°$. These differences are due to the different modes of coalescence in the rock bridge. For $\beta = 105°$ the total failure is preceded by bridge rotation, which appears to take longer than the simpler mode of failure for $\beta = 120°$. It has only been possible to determine the time ranges of interest using the AE signatures, whilst the photographs allow the fracture geometry evolution to be described. Additionally the frequency character of AE events is investigated and shown to have significant potential for characterisation of AE source types and thus failure processes.

**Key words:** Fracture, acoustic emission, digital images, pre-existing flaws, rock bridge, soft rock.

## *Introduction*

The city of Naples (Italy) is built on pyroclastic deposits that originate from the volcanic activity of the Phlegrean Fields and Vesuvio, the two volcanic areas that

[1] Laboratoire Sols Solides Structures, Grenoble, France. E-mail: hall@geo.hmg.inpg.fr
[2] Dipartimento di Ingegneria Geotecnica Università di Napoli Federico II; now at Lombardi Engineering Ltd.

surround the city. These deposits consist of an upper part of cohesionless silty sand (pozzolana) and a lower part of soft rock. The soft rock of the lower layers is the Neapolitan Yellow Tuff, which originated from the lithification of pyroclastic soils produced during the activity of the Phlegrean Fields. Throughout history (at least until the advent of reinforced concrete), the subsurface of Naples has provided an invaluable quarry for the tuff, which makes an ideal construction material since it has good mechanical properties, is quite light and can be easily cut into bricks. In fact, this densely populated city is largely built of such bricks. However, buildings have frequently been built directly above the underground quarries; either to reduce transport costs of the quarried material or because they have been built over forgotten cavities during the city's expansion. These caves have long been abandoned and generally little care is taken to preserve them. Therefore the cavities are frequently in a bad state and local failures often occur leading to the detachment of large blocks from the roofs, walls and pillars, which can induce general collapses. These caves are quite shallow so interaction with structures at the ground level is to be expected. As a result there is a significant hazard of sudden, and usually unexpected, collapse of underground cavities. These collapses can occur a long time after excavation has been completed and over the last 20 years more than 40 catastrophic events in quarries excavated over a hundred years ago have been recorded.

The tuff mass, within which the cavities were excavated, often contains discontinuities of various origins. Some of these discontinuities were generated immediately after the formation of the tuff, due to the cooling of pyroclastic materials, whilst others developed over time around the cavities after excavation and stress redistribution. Visual inspections and borehole investigations have revealed that fractures are quite widely spaced and often not fully persistent in space, i.e., the rock mass around the cavities consists of blocks attached by sections of intact or unbroken segments of rock between fractures (rock bridges). These rock bridges can provide an initial stability to a potential falling block. The delay-time for local collapse after excavation could be explained by gradual deterioration of these rock bridges under the weight of the block. Such progressive failure involves the evolution of non-persistent discontinuities, which serve as stress concentrators. Interaction, propagation and coalescence of the discontinuities leads to the formation of larger fractures and rock bridge failure resulting in block detachment and ultimately the collapse of the entire rock structure (see Fig. 1).

This paper considers the failure processes, which could lead to *in situ* collapse, of the Neapolitan Tuff. The aim of the study is to better understand how and when rock failures occur and what precursory signals may be identified to minimise the effects. An experimental program was developed at Laboratoire Sols, Solides et Structures (3S) in Grenoble to examine the mechanisms of crack initiation, propagation and interaction in natural, dry specimens of Neapolitan Fine-Grained Tuff (FGT). The test series was designed to investigate the failure process of both intact samples and samples with artificial pre-existing flaws (cuts). In the first case the natural

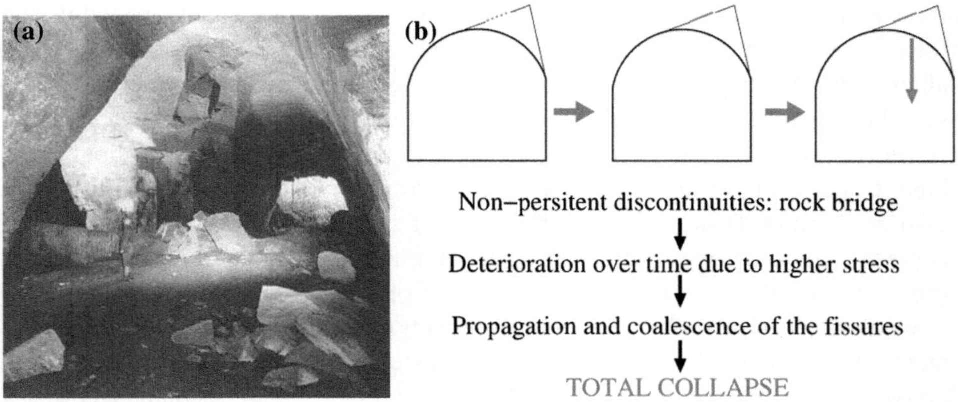


Figure 1
(a) Cavity beneath the city of Naples showing fallen blocks and the resultant exposed fracture surfaces; such block falls can lead to complete collapse of the cavity. (b) The process of block detachment, from the cavity roof, through gradual propagation and coalescence of fractures.

heterogeneity of inclusions, pores and micro-fractures in the volcanic tuff is directly responsible for the micro-mechanical processes that lead to macroscopic axial splitting, under unconfined compression. However, in the presence of pre-existing discontinuities, the cracking process depends on the number and geometry of flaws present and, as will be discussed, involves a number of possible mechanisms. The key focus of this work is understanding these phenomena and their controlling factors. Therefore crack propagation from a single open, inclined flaw and crack coalescence in rock bridges between two open inclined, parallel flaws are studied and compared to the intact specimen tests. Results from the complete experimental program, which investigates the rupture process and coalescence in the rock bridge in the FGT, are discussed in greater detail by DE SANCTIS (2005). The main focus of this paper is the use of acoustic emissions (AE), in conjunction with digital images, to analyse the fracture processes.

### *Fracture Mechanics and Rock Bridge Failure*

Extensive research has been performed on the brittle behaviour of rocks under compression. In particular, with relevance to this current work, experimental studies have previously been conducted on both "rock-like" materials (REYES and EINSTEIN, 1991; SHEN and STEPHANSSON, 1993; SHEN *et al.*, 1995; BOBET and EINSTEIN, 1998; WONG and CHAU, 1998; SCAVIA and CASTELLI, 1998; MUGHIEDA and ALZOUBI, 2004) and real rock specimens (INGRAFFEA and HEUZE, 1980; PETIT and BARQUINS, 1988; JIEFAN *et al.*, 1990) containing two pre-existing inclined flaws through the sample thickness. In all these cases, under uniaxial compression, a common cracking pattern is observed that involves wing and secondary cracks. According to BOBET (2000) these patterns occur through initiation of the wing cracks at the flaw tips followed by their propagation, in a curvilinear path, under increasing load with a tendency to align in the direction of the major principal stress. These are tensile cracks that grow in a stable manner at the outer tips as well as at the inner tips of the flaws. It is often considered that most observed secondary cracks are shear failures that initiate at the inner tips of the flaws and grow in a direction coplanar or nearly coplanar to the pre-existing fracture. However, two other types of secondary cracks are also observed: coplanar cracks propagating from the outer flaw tips and non-coplanar cracks growing parallel to the wing cracks but in the opposite direction (see BOBET, 2000). The latter type of secondary cracks are rarely observed in experiments and seem to be material dependent. In conclusion, despite extensive research and some generally accepted ideas, uncertainties still exist on the nature of these secondary cracks, in part due to the range of different patterns observed. For example, the mechanism of formation of the secondary cracks and if they are pure shear, or if they could be tension induced, are still open questions. Thus efforts are needed to better understand and classify these cracks including defining new experimental methods and testing of other materials.

*Acoustic Emission (AE) Monitoring of Rock Failure and Fracture Development in Laboratory Tests*

Monitoring of acoustic emissions produced in the failure process of rocks has been shown to be very useful in assessing the stages of failure, the locations of active deformation and characterisation of modes of rupture. The method has been successfully applied both in laboratory tests and for *in situ* monitoring. Examples of such AE studies in laboratory rock testing include, among others, SCHOLZ (1968); LOCKNER and BYERLEE (1977); MAIN *et al.* (1989); LOCKNER *et al.* (1992); LOCKNER (1993); LIAKOPOULOU-MORRIS *et al.* (1994); EBERHARDT *et al.* (1998); ZANG *et al.* (1998); LEI *et al.* (2000); AMITRANO (2003); STANCHITS and DRESEN (2003); BAUD *et al.* (2004); LEI *et al.* (2004). The AE technique, at its most basic level, considers the hit-rate of events recorded using a transducer in contact with the test sample or apparatus to provide insight into the timing of fracture processes and, in particular, the onset of failure. Such analysis can be extended to consider amplitude, energy and time-based attributes to further characterise the recorded events and to attempt to classify different types of events and thus monitor changes in the failure mode. Furthermore, the so-called *b value* (originally used to analyse historical earthquake records) has been used to assess AE variations over time (e.g., MAIN *et al.*, 1989; LIAKOPOULOU-MORRIS *et al.*, 1994; ZANG *et al.*, 1998; STANCHITS and DRESEN, 2003; LEI *et al.*, 2004). Such an analysis may provide insight into the type of deformation occurring in a test such as brittle versus ductile failure (e.g., AMITRANO, 2003) or changes from a primary to a nucleation phase (e.g., LEI *et al.*, 2004). More quantitative analysis is possible, with a number of listening transducers, through event location and source mechanism analysis. The former can highlight the transition from distributed to localised deformation that might indicate the onset of macroscopic failure (e.g., LOCKNER *et al.*, 1991; ZANG *et al.*, 1998; LABUZ *et al.*, 2001; STANCHITS and DRESEN, 2003; LEI *et al.*, 2004). Source mechanism analysis, based on received amplitude polarities, provides insight into the failure mode, such as shear versus dilation, and the sense of movement (e.g., ZANG *et al.*, 1998; LEI *et al.*, 2000). This can be extended, with appropriate data, to invert for the focal mechanism of the source (e.g., CARVALHO and LABUZ, 2002; COLLINS *et al.*, 2002). Such analysis can thus help the understanding of mechanisms active in rock failure.

### *Neapolitan Fine Grained Tuff (FGT)*

The Neapolitan FGT is a natural pyroclastic rock that resulted from deposition, approximately 15,000 years ago, and diagenesis of hot pyroclastic products. The structure of FGT consists of an ashy matrix with high porosity (pumice) and lithic inclusions embedded in a zeolitic cement (mainly phillipsite and cabasite minerals). The inclusions are randomly arranged and their size is usually smaller than 1 mm, although pumice fragments of a few centimetres are also found. The average physical and mechanical properties of FGT, given by AVERSA *et al.* (1991) and AVERSA and EVANGELISTA (1998), are summarised in the following. The FGT has a high porosity (ranging between 0.4 and 0.5), as is typical of pyroclastic rocks, that is made up of pores with an average size of between $8 \times 10^{-4}$ and $2 \times 10^{-3}$ mm, as determined by mercury porosimetry. The uniaxial compression strength of dry FGT falls in the range proposed as lower and upper limits for soft rocks (1.2–25 MPa: Associazione Geotecnica Italiana, 1977; Geological Society Engineering Group working party, 1970). This class of geomaterials exhibits mechanical behaviour between those of soils and hard rocks depending on the stress state. In the case of the FGT, the material behaviour changes from “rock-like” to “soil-like” as confining stress increases, with a transition at about 7 MPa.

## *Experimental Procedure*

### *General*

The experiments presented in this work were performed in the plane-strain compression (biaxial, hereafter) apparatus developed in Grenoble by DESRUES (1984) and later modified by others (see DESRUES and VIGGIANI, 2004). This apparatus has two key advantages for this study: 1) fracturing can be studied in 2-D, because plane-strain conditions are, in general, ensured; 2) crack patterns can be tracked during the test, because visual inspections of the sample faces are possible. Figure 2 shows the principal elements of the test rig. The device consists

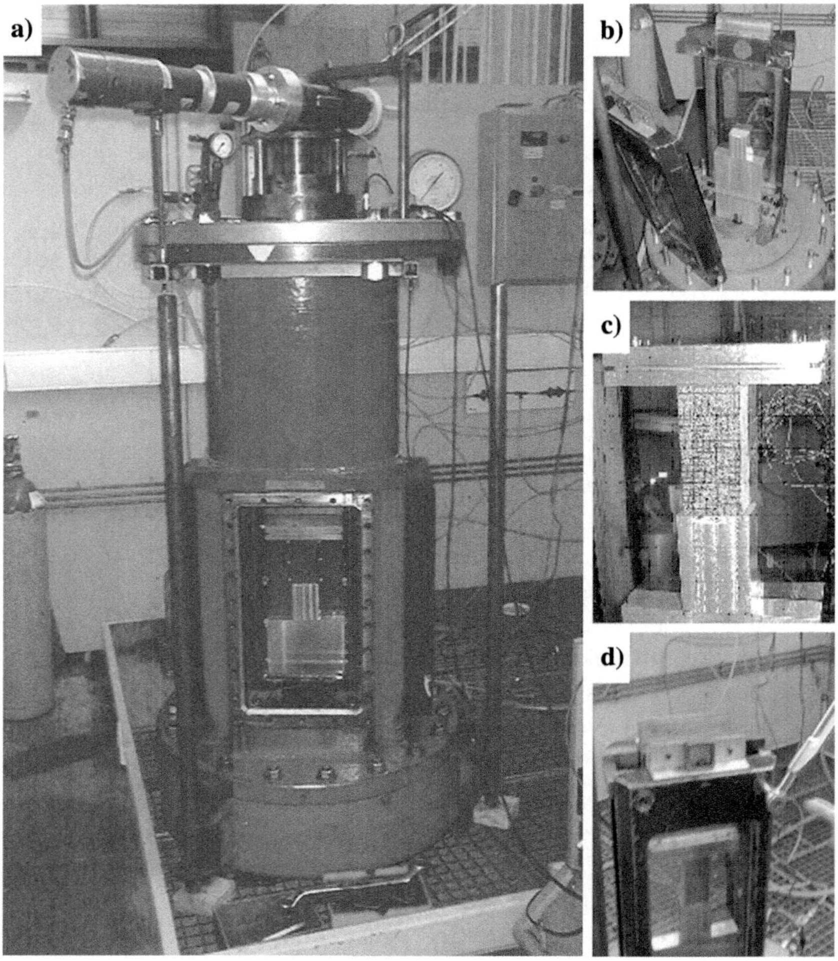


Figure 2
The Laboratoire 3S biaxial apparatus and experimental set-up. The device, shown in its fully assembled state in (a), consists of two parts: the fixed lower base (b) and the mobile upper cell. Prismatic specimens are mounted between two 50 mm thick, rigid, parallel glass platens, which induce plane-strain conditions (b, c, d). The large portholes and see-through glass platens (a, d) allow observation of the specimen during a test. Loading is applied by a displacement-controlled jack that can be seen at the top of the apparatus in (a).

of two parts: the fixed lower base (Fig. 2a) and the mobile upper cell (Fig. 2b). The prismatic specimen is mounted between two 50 mm thick, rigid, parallel glass platens, which induce plane-strain conditions. All the surfaces in contact with the specimen are lubricated to minimise friction. Although the apparatus allows for saturated specimens and the application of a confining fluid pressure, these tests are under dry and unconfined conditions. The cell has large portholes allowing observation (and the facility to take photographs) of the larger (in-plane) faces of the specimen during the test, and thus crack propagation can be traced. Loading is applied by a displacement-controlled jack at a constant rate of 3 $\mu$m/min in the vertical direction; this displacement is measured, along with the applied load, throughout a test until failure is reached. In this work, the experimental set-up also included monitoring AE with transducers attached directly to the unconstrained sides of the specimens, as discussed below.

*Specimen Description and Preparation*

The FGT used in this study was acquired in irregular blocks resulting from road maintenance works around outcrops in the centre of Naples. Prismatic test samples were cut from these blocks using a rock saw; each sample had dimensions of: 100 mm high, 50 mm wide and 35 mm thick (see Fig. 3). Specimen thickness is dictated by the distance between the glass platens of biaxial apparatus. The chosen width of the specimens of 50 mm is thought to be sufficient relative to the matrix grain size and is also in the range of values reported in the literature for similar studies. Pre-existing flaws were made in the samples, perpendicular to the larger faces, using a diamond wire saw (at the CEA in Grenoble). For the rock bridge studies two flaws were made but samples with a single flaw or no flaws were also tested as the base-case scenarios. The flaw parameters were: length = 12 mm; aperture = 0.4 mm; inclination = 45° (see Fig. 3). A range of flaw configurations was created by varying the rock bridge angle (the angle between the inner-tips of the two flaws, hereafter denoted $\beta$) from

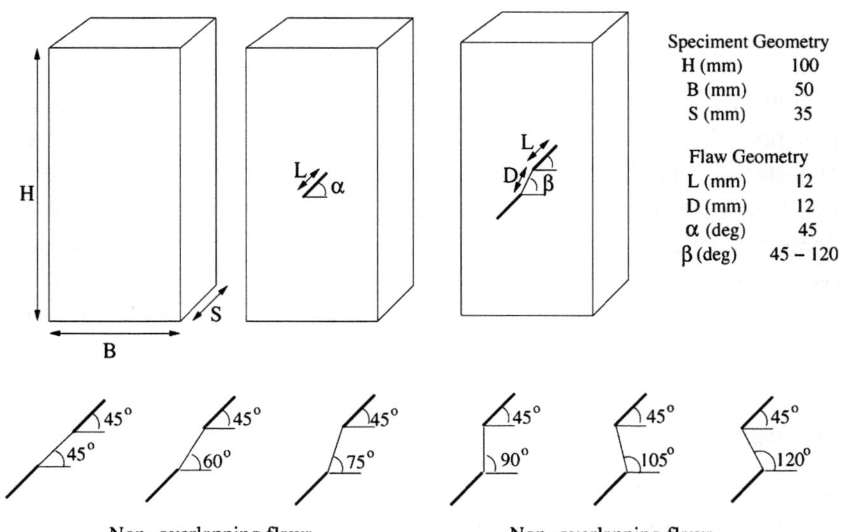


Figure 3
Diagram of the prismatic test samples and key dimensions plus a summary of the flaw configurations (differentiated by the angle, $\beta$, between the inner tips of the flaws) investigated in the testing program. This paper considers primarily the cases for $\beta = 105°$ and $\beta = 120°$.

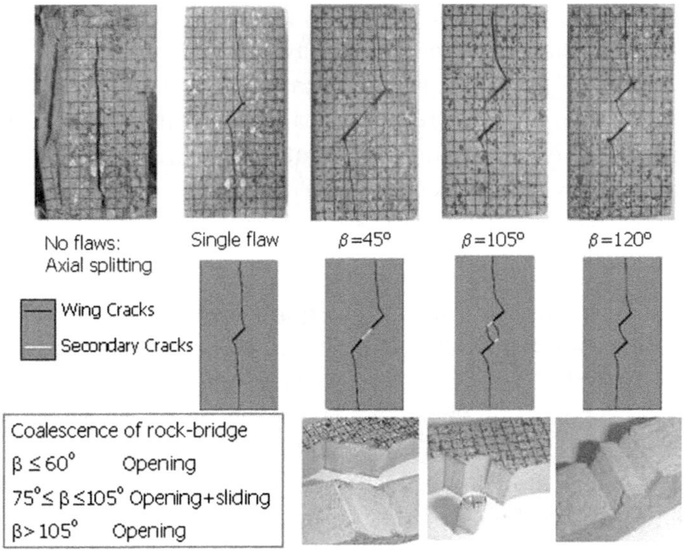


Figure 4
Summary and examples of flaw configurations, final fracture patterns and rupture modes. The top line of photographs show example specimens after testing. Note in particular: the axial splitting when there are no pre-existing flaws; the opening mode failure with classic wing cracks for a single flaw; wing cracks and secondary shear coalescence across the rock bridge for $\beta = 45°$; the creation of a rock cylinder through inner wing crack coalescence, rotation and secondary shear cracking in the rock bridge for $\beta = 105°$; the more simple wing crack coalescence for $\beta = 120°$. The nature of the failure in each case is summarised in the table, based on evidence from the photographic data and observations of the rupture surfaces after the tests (lower row of photographs).

45° to 120° with 15° increments (Fig. 3); other dimensions were kept constant. Figure 4 provides a summary of the flaw configurations studied and the observed failure patterns. A key feature of the sample failure mechanisms is the dependence of the rock-bridge mechanics on the angle $\beta$. For example, DE SANCTIS (2005) shows that there is a clear trend in peak stress with changing $\beta$: the lowest strength is observed around $\beta = 90°$, which marks the change from non-overlapping to overlapping flaws (Fig. 3) and the transition between two different coalescence mechanisms.

### *Digital Images*

The Laboratoire 3S biaxial set-up allows direct observation of the specimen faces through the transparent platens and the portholes (see Fig. 2). Various methods of visual monitoring of the deformation plane and tracing of crack pattern development were assessed by DE SANCTIS (2005) who found that the best results are obtained with a high-resolution digital camera (in this case a Kodak professional DCS Pro 14n and 13.5 mega-pixel images). With such photographs the development and propagation

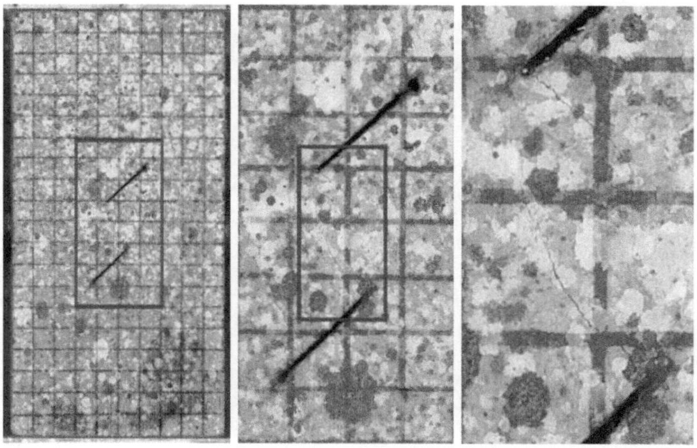

Figure 5
An example of the detail in the digital photographs showing the development of very small fissures across the rock bridge.

of fracturing can be mapped, including small details as in Figure 5. There is also the advantage that the images can be analysed in detail and at different scales (through zooming) after the test and with no need to stop the loading, e.g., to identify and focus on emerging cracks.

*AE Monitoring*

Two AE transducer set-ups were used for the AE monitoring in conjunction with a MISTRAS2001 acquisition system and standard preamplifiers (Physical Acoustic Corporation, 1220A). The two set-ups are shown in Figure 6 with the transducers in direct contact, aside from the couplant, with the unconstrained sides of the specimen. Results from four different tests for each of the set-up configurations are presented later.

The first set-up (set-up I) used two transducers (Digital Wave Corporation, B1025, diameter = 10 mm) that had a good broad-band frequency response, over the range of approximately 50 kHz – 2 MHz, which allowed spectral analysis of the waveforms, as discussed later. These transducers were attached symmetrically on the free sides of the specimens using silicon vacuum grease as a coupling agent. The second AE set-up (set-up II) used four transducers (Physical Acoustic Corporation, R15, diameter = 17 mm), which allowed event location to be performed. In this case, two transducers are symmetrically mounted on each side of the specimen (Fig. 6) using Sofranel SWC couplant and silicon to flexibly bond the transducers to the sample so that the effects, on the coupling, of any change in shape of the sample was minimised. This improved transducer-sample coupling procedure provided a better energy response and provided additional shielding from external noise. However, these transducers have a poor frequency response in that they are not

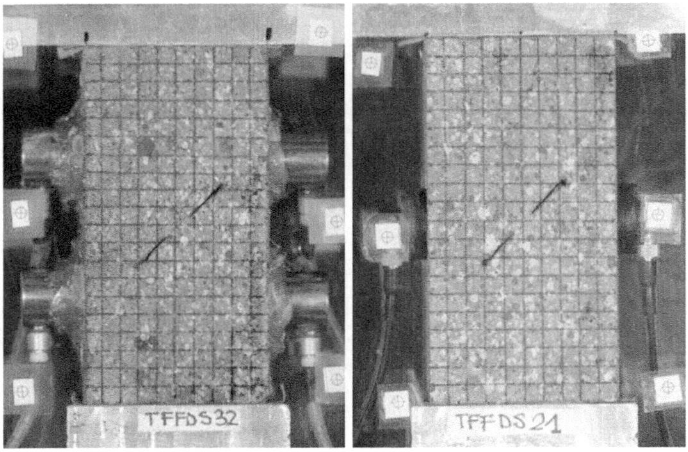


Figure 6
Photographs of the AE transducer set-ups for two tests just prior to loading. The AE transducers are placed near the centre to focus on the activity around the rock bridge area. Also note the grid that is drawn onto the samples prior to testing and the black and white speckles (paint sprayed onto the surfaces), both of which aid the observation of displacements and rupturing in the photographs. For set-up II (left), with four transducers, transducer 1 is top left, 2 top right, 3 bottom left and 4 bottom right.

broad-band (main operating frequency range is about 100–600 kHz) and have significant resonance. Therefore spectral analysis was not carried out for the tests using these transducers (it may be possible in the future to deconvolve the transducer response to ameliorate the signal but currently this has not yielded good results due to a lack of good characterisation of the response function).

In both set-ups, signals from the AE transducers were pre-amplified by 40 dB gain and passed through a frequency filter (100–1200 kHz) and a post-amplifier of 40 dB gain inside the recording system. In each case full waveform data were recorded for each event with a sampling rate of 4 MHz and 1024 points. The system settings were kept constant for all the tests except the trigger threshold, which was adjusted within the range 30–45 dB to achieve the best monitoring depending on the specific situations (noise, etc.) and variations in coupling between the transducers and the specimen (determined before each test using the standard "pencil-break" test). Other settings were: Peak Definition Time = 200 $\mu$s; Hit Definition Time = 800 $\mu$s; Hit Lockout Time = 1000 $\mu$s (see Physical Acoustics Corporation, 2001, for details).

## *AE Analysis*

### *AE Attributes*

For both the AE set-ups, various AE attributes were assessed for insight into failure modes and precursory signals to specimen failure. The key attributes considered were AE counts and AE energy. Other attributes, such as average frequency, event amplitude, rise time, etc., were not found to provide significant insight in this work. Additionally a number of advanced techniques were used to characterise the AE recording and provide better understanding of the failure development. These methods are described below for the two set-ups separately since different analyses were possible in each case.

*Spectral Analysis (Set-up I):*

With broad-band transducers it is possible to perform spectral analysis of full waveform data. This involves looking at the fast Fourier transform (FFT) of each time-series waveform (here 1024 points) and assessing the distribution of energy across the frequency range. This spectral analysis was carried out to investigate if it could provide improved characterisation and differentiation of event types (e.g., shear versus opening-mode) and identification of different phases of failure. In particular it was hoped that better precursory signals to failure might be identified.

*Event Location and $^{\Sigma N}/_{\Sigma E}$ Analysis (Set-up II):*

In the second set-up four transducers were used, which allowed 2-D event location to be carried out. A 2-D approach is appropriate in this work since the tests are under plane-strain conditions and the ruptures are generally seen to be roughly perpendicular to the main faces of the sample, i.e., the failure is in effect 2-D. The 3-D width of the sample will result in some smearing of the events when presented on a 2-D figure but this effect is minimal. The location of events was performed by first grouping the events into four spatial sectors, based on their arrival times at each transducer, and selecting the three nearest transducers (i.e., those with the earliest arrival times) to carry out the location inversion. Onset times were given by the MISTRAS2001 automatic trigger. Using just three transducers for the inversion reduced the error due to picking inaccuracy, which increases with the travel time. However, only events that were recorded on all four transducers simultaneously were used, ensuring that only the stronger events were considered; smaller events again have higher picking errors due to difficulty in defining the true onset. Thus, for each chosen event and the three selected transducers, the best-fit solution for the event position was determined, using the differential arrival times, $dt_2 = t_2 - t_1$ and $dt_3 = t_3 - t_1$, through iteration to minimise the functions,

$$
\begin{aligned}
F_2 &= 2R(dt_2 v + D_2 \cos(\phi - \theta_2)) - D_2^2 + dt_2^2 v^2, \\
F_3 &= 2R(dt_3 v + D_3 \cos(\theta_3 - \phi)) - D_3^2 + dt_3^2 v^2,
\end{aligned} \tag{1}
$$

(after, e.g., GLEA, 2002) where $F_{2,3}$ are equal to zero for the perfect solution and otherwise are minimised for the best-fit solution, $R$ and $\phi$ are the distance and angle of the event from transducer 1, $D_{2,3}$ and $\theta_{2,3}$ are the distances and angles to transducers 2 and 3 from transducer 1 and $v$ is the wave velocity in the medium. The velocity in the sample will evolve with loading and will vary spatially, so ideally, some form of initial characterisation of the velocity field before the test and tracking of changes through the test should be carried. This was not done in these tests. A best-fit constant velocity was determined for each test that gave the best localisation and corresponds approximately to the shear-wave velocity of the FGT (in the range 700–800 m/s), since the MISTRAS2001 trigger was based on the large amplitude (shear-wave) arrival and not the first (P-wave) onset.

As discussed earlier, AE "b values" have previously been used to assess AE events and their occurrence over time. For example, the b value (given by the second constant in the empirical relationship of the number of events, N, of a size greater than or equal to magnitude $M$, $\log_{10}N = a-bM$,) is often seen to drop during the enlargement of cracks (e.g., SHIOTANI *et al.*, 2000). Such empirical analysis can also be applied to acoustic emission data, after multiplication by 20 to correct for the fact that AE amplitudes are measured in decibels rather than the logarithmic peak amplitude of the Richter scale (HATTON *et al.*, 1993). Typically the b value is around 1 but higher values are often quoted for AE, e.g., 2.4 for sandstone and 1.4–3.8 for granites (HAZZARD and YOUNG, 2002). In this work, values of 2–6 were observed for the FGT, indicating that for soft rocks higher b values might be expected. The Appendix presents some example b value plots and highlights some issues in the analysis.

The ratio, ${}^{\Sigma N}/_{\Sigma E}$, of the cumulative hits, $\Sigma N$, and the cumulative energy, $\Sigma E$, is used in this work instead of the b value (the inverse of this parameter —${}^{\Sigma N}/_{\Sigma E}$— was used by ZANG *et al.*, 1996). As will be shown, ${}^{\Sigma N}/_{\Sigma E}$, which might be thought of as an inverse average energy parameter, provides an effective and informative indicator of changes in the deformation phases throughout the tests for the FGT. This ratio allows combined interpretation of the two first-order AE parameters (hits and energy) without the need for line-fitting or windowing (although it is noted that there is an inherent windowing over the whole test time); the Appendix presents a comparison of some typical b value and ${}^{\Sigma N}/_{\Sigma E}$ plots for this test campaign and highlights some issues. The hypotheses for interpretation of the ${}^{\Sigma N}/_{\Sigma E}$ values are that an increase indicates the occurrence of numerous small, low-energy events, which could be associated with fracture propagation or shear fracturing, and a decrease indicates a small number of high energy events, which may occur at fracture initiation and particularly for opening mode ruptures.

## *Results*

A thorough analysis of the experimental program results both with and without AE monitoring has been presented by DE SANCTIS (2005), as summarised in Figure 4. In this section the specifics of the AE results and their correlation with the rupture process, as observed from load parameters and the high resolution photography, are discussed in detail. Firstly, results from the studies involving no pre-existing flaws or just a single flaw are presented as these represent the "base cases" and are expected to highlight the effects of sample heterogeneity and simple rupture modes; in these cases only opening mode fracturing is expected. These tests are discussed with respect to a classical interpretation of fracture development, which appears appropriate. Following these base cases, results are presented for key tests from the rock-bridge studies, i.e., those with two pre-existing flaws. In particular tests involving $\beta = 120°$ and $\beta = 105°$ are shown. In each case, two primary AE characteristics, cumulative hits and cumulative energy, and the ratio of these values $^{\Sigma N}/_{\Sigma E}$ are presented. The analysis of the variations of these characteristics as a function of time provides information on the chronology of the processes, which can be compared with the axial stress evolution, as monitored simultaneously throughout the test. The different phases of the deformation processes are determined primarily from the $^{\Sigma N}/_{\Sigma E}$ trends. Additionally AE event locations are presented both as points, with size proportional to the relative event magnitudes, and spatial number-density contours. These results are discussed in a following section with a comparison to the more classical failure observed for the base cases. Subsequently analysis of the spectral signatures of the AE recordings is presented for key tests (no cuts, $\beta = 45°$, $\beta = 105°$ and $\beta = 120°$) using set-up I, which had just two broad-band transducers. The aim of the presentation of these latter results is to demonstrate what additional insight might be gained by considering the full depth of the AE data. Unfortunately, since only two transducers were used in this set-up, the results cannot be tied into event location. Even if not all data are available for every test due to technical problems, a general consistency between the results is maintained.

### *No Flaws* (*Tffds35*): *Figures 7, 8 and 9*

*Crack pattern evolution*: Figure 7 shows the final state of sample *Tffds35*, which had no pre-existing artificial flaws, after it was removed from the apparatus. From this photograph it can be seen that the fracturing involved just axial splitting, i.e., opening mode fractures, and that the failure was not 2-D—two principal fractures are seen that form a rough "V" in horizontal cross section across the sample axis with roughly vertical traces at the front and a vertical "Y" at the back. These two macro-fractures were only visible on the front face at the end of the test, whilst they could be seen much earlier at the back, therefore the photos from during the test unfortunately provided little information. For this test the natural heterogeneity of the material has to be considered the main cause of crack initiation.

*Acoustic emissions*: Based on the stress-time curve, the failure process can be classically interpreted as having the following stages (e.g., after BIENIAWSKI, 1967): 1. Crack closure; 2. Linear stress-strain response and crack initiation with stable crack growth; 3. Crack damage and unstable crack growth leading to failure. AE allows these phases of rupture to be better identified and characterised:

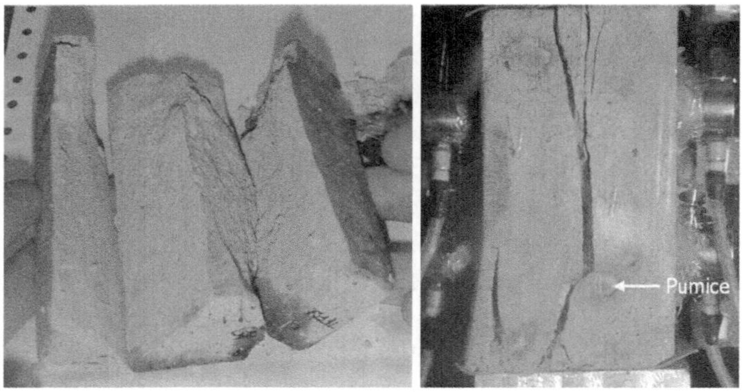


Figure 7
Photographs of sample *Tffds35*: left — after removal from the test cell at the end of the test to show the final rupture achieved with the top of the sample to the fore; right — the back of the sample at the end of the test. This sample had no pre-existing flaws and failure was through axial splitting with cracks initiating at the pumice inclusion highlighted. It can be seen that two main rupture planes formed that are not parallel to the unconstrained sides of the sample so a plane strain condition was not maintained.

- Phase 1: During the crack-closure phase the stress-time curve increases non-linearly with an increasing axial stiffness (Fig. 8). During this phase little AE activity is recorded and event locations are randomly distributed over the sample, which reflects the heterogeneity of the material (Fig. 9). Also the ${}^{\Sigma N}/_{\Sigma E}$ ratio is seen to increase until it reaches a plateau at about 1 hour, which presumably indicates that the majority of existing cracks closed early on.
- Phase 2: With increasing load the stress-time curve becomes linear corresponding to an increase in the gradient of the cumulative hits, energy and ${}^{\Sigma N}/_{\Sigma E}$ curves at about 2.9 hours. Initially the AE events are focused around a large inclusion (Fig. 9) but an alignment of events is also seen in the upper part of the sample with a prevalent vertical distribution and a "Y" that resembles the final fracture pattern to the rear. The emergence of this alignment of AE events is taken to indicate the macro-crack initiation. The higher level of AE activity continues until failure.
- Phase 3: The unstable crack phase is characterised by a non-linear stress-strain response with a decrease in axial stiffness (Fig. 8) and a significant drop in ${}^{\Sigma N}/_{\Sigma E}$ at about 8.06 hours. After this non-linear phase, two peaks are seen in the stress-time curve at about 8.4 and 9.2 hours, each associated with large jumps in both the AE energy and counts (Figs. 8a, b). These features indicate the complete development of the macro-cracks, which probably involves propagation of the

►

Figure 8
AE statistics for *Tffds35* (no flaws) versus time over the duration of the test: (a) cumulative AE hits for the four transducers and the loading curve (note strain is proportional to time); (b) cumulative AE energy for the four transducers and the loading curve; (c) ${}^{\Sigma N}/_{\Sigma E}$ for the four transducers plus the loading curve. The periods of interpreted deformation phase, as described in the text, are shown.

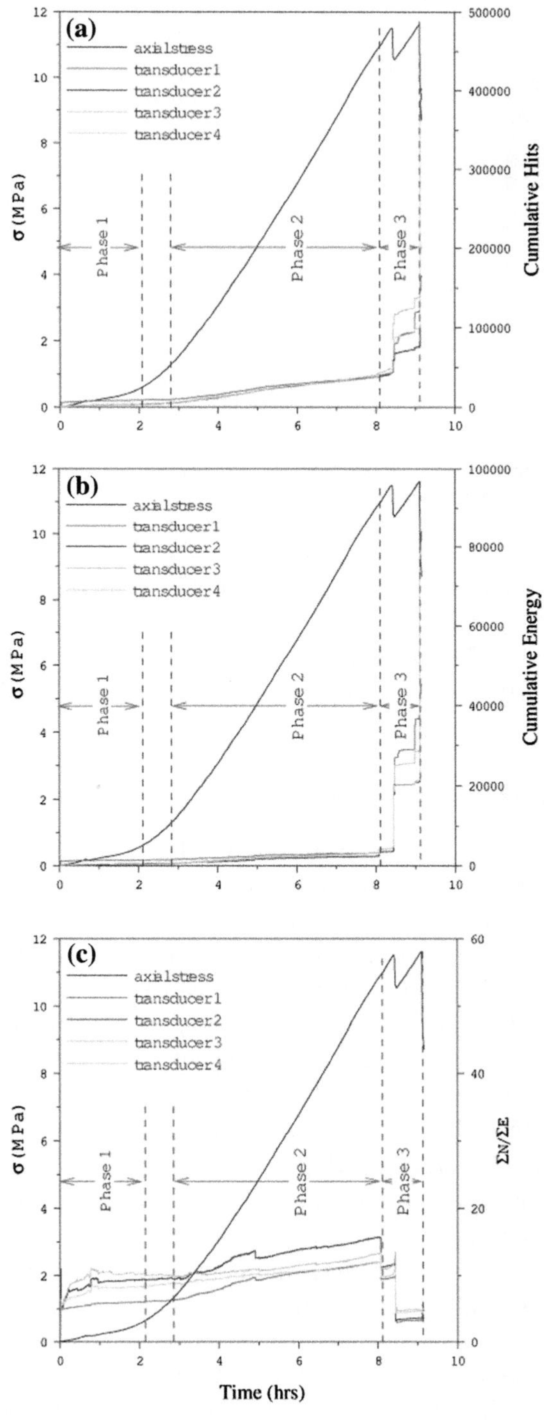
(a)
axialstress
transducer1
transducer2
transducer3
transducer4
σ (MPa)
Cumulative Hits
Phase 1
Phase 2
Phase 3
(b)
axialstress
transducer1
transducer2
transducer3
transducer4
σ (MPa)
Cumulative Energy
Phase 1
Phase 2
Phase 3
(c)
axialstress
transducer1
transducer2
transducer3
transducer4
σ (MPa)
ΣN/ΣE
Phase 1
Phase 2
Phase 3
Time (hrs)

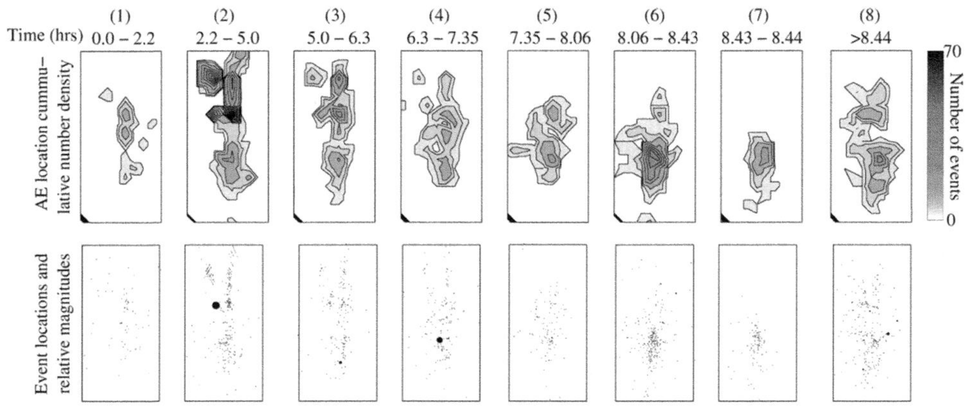


Figure 9
AE event locations for test *Tffds35*. Upper plots show the contoured spatial number density of events during each time window, whilst the lower plots show the actual event locations and the relative event magnitudes (indicated by the symbol size).

cracks from the back to the front of the sample, giving the horizontal broadening of the AE event location distribution (due to the projection of a 3-D distribution onto the 2-D plane). The final sample failure occurs after a period of activity at a few points along the fracture surface accompanied by a small increase in $^{\Sigma N}/_{\Sigma E}$ Fig. 8c).

*Single Flaw (Tffds33): Figures 10 and 11*

*Crack pattern evolution*: Figure 11 shows the fracture development for the single-flaw test *Tffds33*, as determined from the photographs taken during the test. The failure mode of this specimen is similar to the case with no flaws, i.e., purely opening, and so this type of test, and the case with no flaws, acts as a control for the characterisation of opening versus shear fracturing. In this case the rupture process involves both wing cracks and secondary tensile cracks which start at the flaw tips. Once initiated, the wing cracks propagate out to traverse the full sample height at failure. In many of the tests with this configuration two secondary cracks (so-called because they develop after the wing cracks) initiated at the same tips before the complete wing-crack development. For this test the mechanism of failure was not fully planar as seen in the different traces on the front and back faces of the wing crack at the upper tip.

*Acoustic emissions*: The stress-time, cumulative hit, cumulative energy and $^{\Sigma N}/_{\Sigma E}$ curves (Fig. 10) show similar trends as in the previous test and the same three phases can be identified:

- Phase 1: The presence of the flaw does not appear to change the crack closure phase, confirming that the flaw aperture (0.4 mm) is sufficiently large to remain open during the whole test.

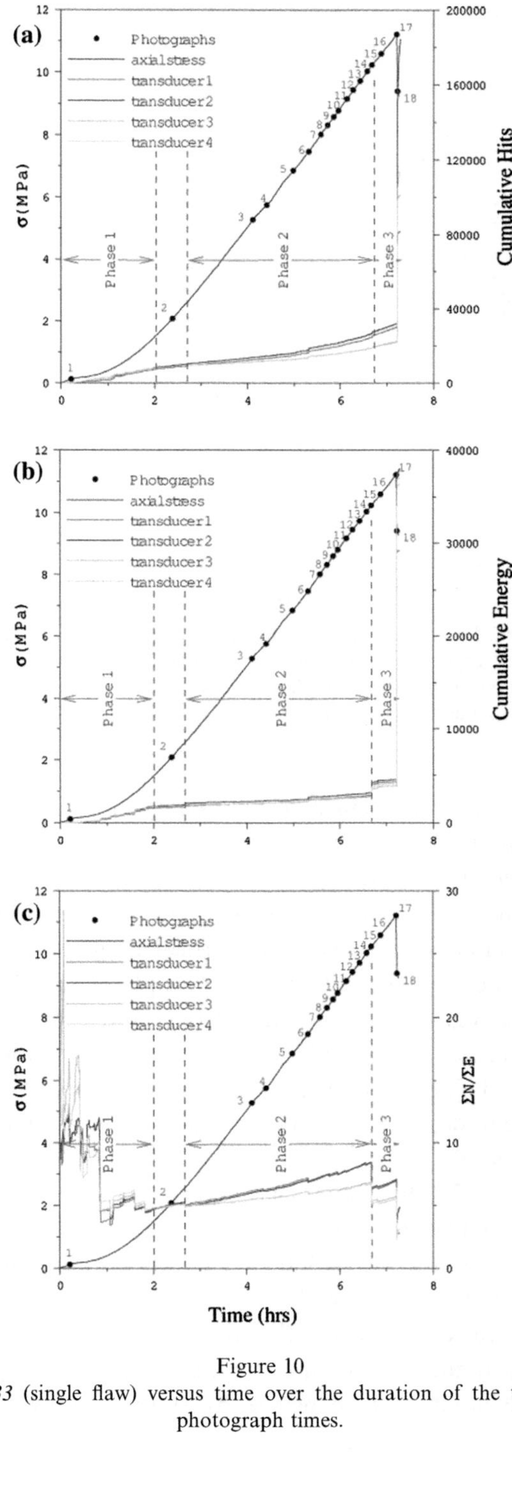


Figure 10
AE statistics for *Tffds33* (single flaw) versus time over the duration of the test as in Figure 8, plus photograph times.

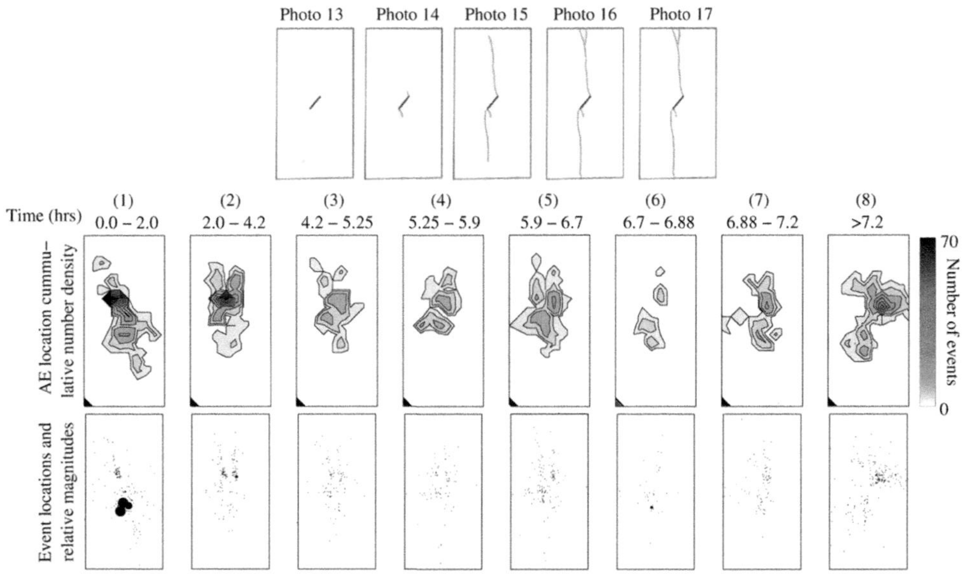


Figure 11
Fracture development and AE event locations for test *Tffds33* (single flaw). The upper row of figures shows the fracture development as mapped from the photographs, taken during the test through the cell's sidewalls, at the times indicated in Figure 10. The lower two rows of figures show the AE event location as contoured spatial number density and location-magnitude distribution, as in Figure 9.

- Phase 2: The onset of two tensile wing cracks at the two flaw tips (as observed later in the photographs) is suggested by the AE locations between 4.2 and 5.25 hours (AE panel 3) and slight drop in $^{\Sigma N}/_{\Sigma E}$ at about 4.2 hours. This is not obvious in the photographs until later (photograph 14 at about 6.5 hours). Before this time the events show some near vertical alignment, which is not clearly associated with the final crack positions. This phase actually appears to start at about 3.6 hours, marked by an increased gradient of the $^{\Sigma N}/_{\Sigma E}$ curve. The subsequent progressive opening of the upper crack, before the lower one, might be indicated by the higher activity recorded on the upper transducers (1 and 2) (Fig. 10) and event locations (Fig. 11), whilst there is little increment for the lower transducers (3 and 4).
- Phase 3: At about 6.7 hours there is clearly activity focussed on the crack tips as a first drop in the load curve and $^{\Sigma N}/_{\Sigma E}$ ratio is seen. This also seems, from the photographs, to be associated with the formation of the secondary tensile crack in the lower part of the specimen, with closure of the wing crack at the lower tip, and opening of both crack apertures so that they are visible on the photographs. As in the *Tffds35* test, at the peak of the stress-time curve a drop in the $^{\Sigma N}/_{\Sigma E}$ ratio indicates sample failure, which occurs by rupture along the two main cracks across the length of the sample.

*Rock Bridge Studies*

$\beta = 120°$ *(Tffds34): Figures 12, 13 and 14*

*Crack pattern evolution*: Figure 12 shows the flaw and transducer configurations for the test *Tffds34*, with two flaws and $\beta = 120°$, plus the final rupture pattern, which involved both external wing cracks and coalesced wing cracks between the lower and upper inner-tips across the bridge. From the photographs (Fig. 14) it appears that cracks initiated at all four tips at the same time, but the external wings developed more rapidly than the inner ones. When the external wing cracks propagated significantly towards the boundaries of the sample, the internal cracks connected across the rock bridge and final sample failure occurred.

*Acoustic emissions*: From the AE data (Fig. 13) three main phases can be identified that are not clear from the load curves or photographs, these are:

- Phase 1: is the settling stage with high, but varying, values of ${}^{\Sigma N}/_{\Sigma E}$ and AE activity concentrated in a couple of sites away from the final fracture positions. The sample is seen, from the ${}^{\Sigma N}/_{\Sigma E}$ curves, to stabilise abruptly at about 1.6 hours and before the start of Phase 2.
- Phase 2: starts at about 2 hours with an increase in the gradient of the ${}^{\Sigma N}/_{\Sigma E}$ curves. Between 1.8 and 2.7 hours the AE event locations are aligned in the lower part of the sample and higher hit rates are seen on the two lower transducers (3 and 4), which could be associated with initiation and propagation of the lower wing crack (Fig. 14). However, no trace of fracturing is seen in the photographs until photograph 3 at about 2.8 hours. Some initial fracturing occurred mainly at

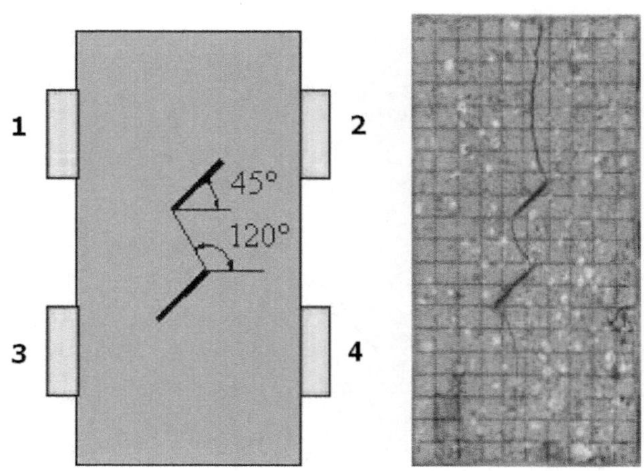


Figure 12
Flaw configuration and transducer arrangement (left) for test *Tffds34* ($\beta = 120°$) plus a photograph of the sample after the test showing the final rupture pattern (the grid drawn on the sample has squares of approximately 5 × 5 mm.

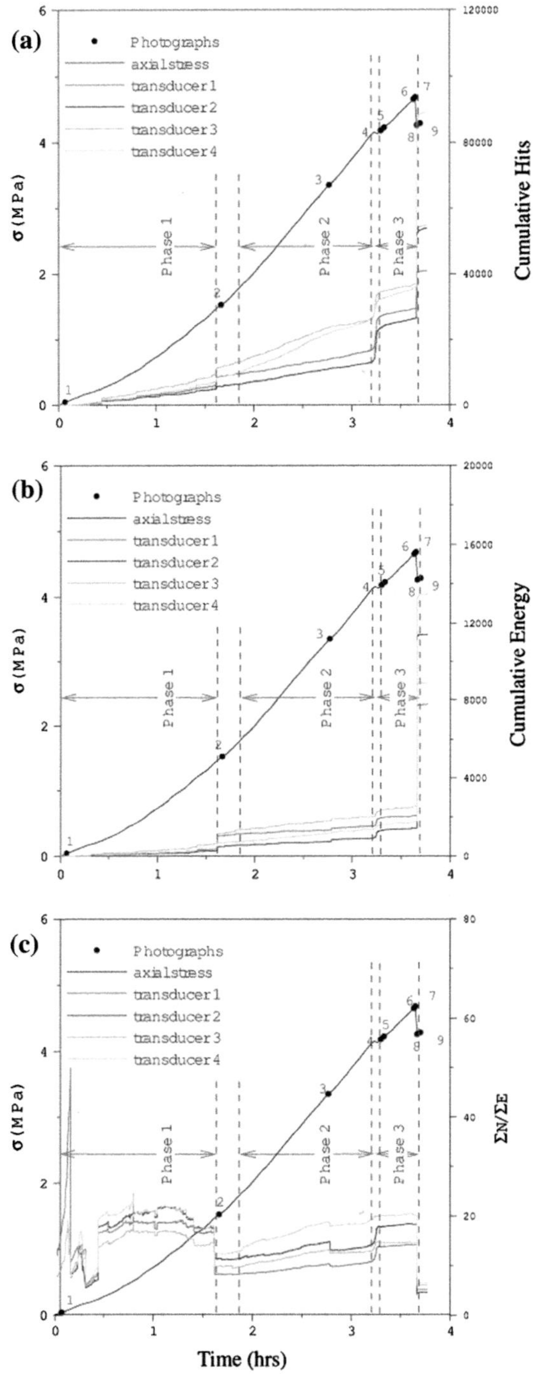


Figure 13

AE statistics for *Tffds34* ($\beta = 120°$) versus time over the duration of the test as in Figures 8 and 10.

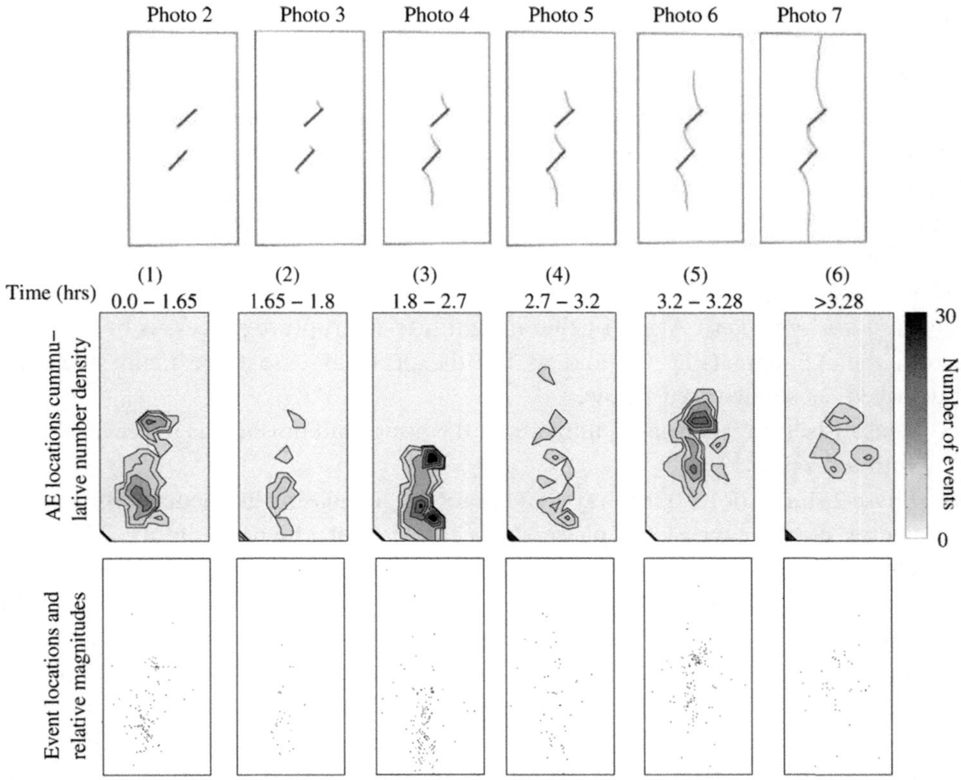


Figure 14
Fracture development and AE event locations for test *Tffds34* (with $\beta = 120°$). The upper row of figures shows the fracture development from the photographs, as in Figure 11, at the times indicated in Figure 13. The lower two rows of figures show the AE event location, as in Figures 9 and 11.

the rear of the sample and so was not seen, but a drop in $^{\Sigma N}/_{\Sigma E}$ at about 2.8 hours may indicate the opening of the wing cracks near the tips, which allowed them to be visible in photograph 3.

- Phase 3: starts at about 3.25 hours with a jump in $^{\Sigma N}/_{\Sigma E}$ (plus increases in both energy and counts) and events located around the rock bridge and lower wing crack position. The location of the AE events (panel 5 in Fig. 14) plus observed cracks in photographs 4 and 5, suggests this initial part of phase 3 is associated with the development of the inner wing cracks. The AE activity stabilised for a short time until failure at about 3.75 hours. At this time the AE activity showed a large increase in both energy and counts but a reduction in $^{\Sigma N}/_{\Sigma E}$ with event locations near the centre of the sample (although it should be noted that exact location of events is difficult at this stage in the test). The AE and photographs suggest that the rock bridge coalescence just preceded sample failure, which occurred through the final propagation of the outer wing cracks to the edges of the sample.

*$\beta = 105°$ (Tffds31): Figures 15, 16 and 17*

*Crack pattern evolution*: Figure 15 shows the transducers' configuration and final fracture pattern for test *Tffds31*. No photographs are available from during the test, however in other tests of this configuration it has been observed that the wing cracks originate at the inner and outer flaw-tips followed by propagation of the external cracks (DE SANCTIS, 2005). Final rupture is normally preceded by connection and rotation of an ellipsoidal rock cylinder within the rock bridge; as can be seen for a similar test sample in Figure 4.

*Acoustic emissions*: Again further insight into the rupture process is to be gained from the AE data (Fig. 16) and as for the previous case three main stages are identified, as summarised below.

- Phase 1 (settling) continues until about 0.6 hours and finishes, as in previous tests, with a drop in $^{\Sigma N}/_{\Sigma E}$.
- Phase 2 starts soon after the system has settled; increasing gradients in the hits and energy curves suggest the phase starts properly at about 1.1 hours. The phase continues until about 2.4 hours, during which time the event locations (Fig. 17) are mainly around the flaws but with some extension upward and downwards (AE panel 2 in Fig. 17). This activity is probably associated with the four wing cracks initiating at the flaw tips followed by some propagation of the external ones.
- Phase 3 has two clear stages, from about 2.4 to 2.55 hours and from 2.55 to about 3.56 hours. The first stage is marked by a jump in $^{\Sigma N}/_{\Sigma E}$ with AE activity focused at the centre, within the rock bridge (AE panel 3, Fig. 17). Through this stage there is a gradual increase in $^{\Sigma N}/_{\Sigma E}$ until a large jump at about 2.55 hours when there is also a step in the loading curve. The AE rate is seen to reduce gradually and the sample stiffness is seen to recover although another couple of

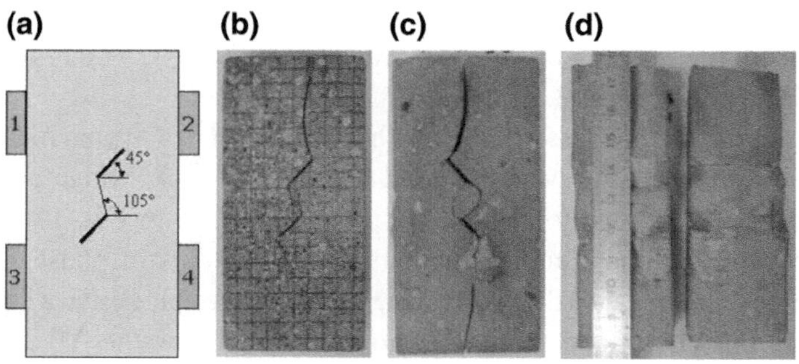


Figure 15
Flaw configuration and transducer arrangement (a) for test *Tffds31* ($\beta = 105°$) plus photos of the (b) front and (c) back of the sample after the test showing the final rupture pattern (the grid drawn on the sample again has squares of approximately 5 × 5 mm). Note the ellipsoidal cylinder that has formed in the rock bridge. (d) shows the sample after removal from the test apparatus highlighting the large inclusion that caused the slight non-planarity of the failure and which was also the site of significant AE activity, as seen in Figure 17.

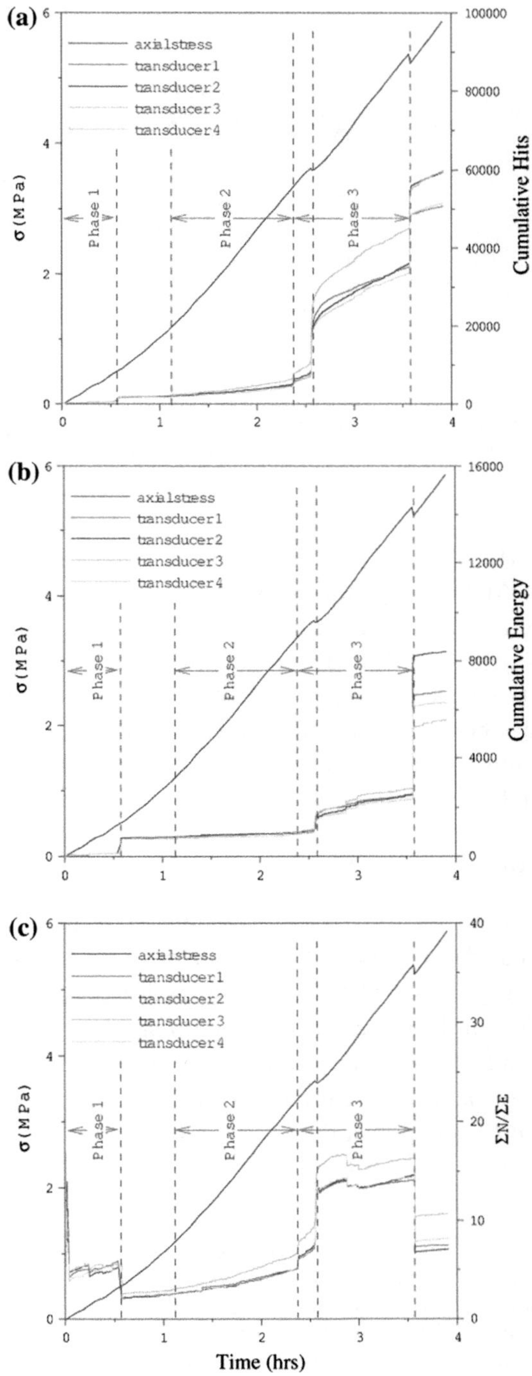


Figure 16
AE statistics for *Tffds31* ($\beta = 105°$) versus time over the duration of the test as in Figures 8–13.

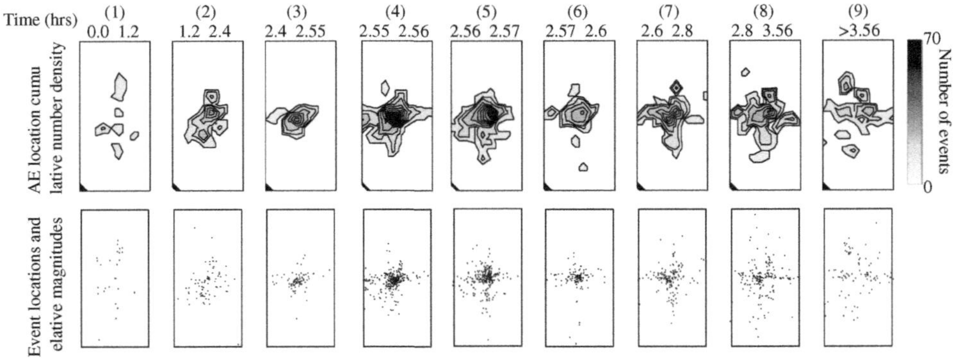


Figure 17
AE event location for *Tffds31* ($\beta = 105°$), as in Figures 9–14. Unfortunately no photographs from during the test are available in this case but the results can be compared with the final rupture pattern in Figure 15.

small steps in AE are seen between about 2.75 and 3.0 hours as the system stabilises again. At the transition between the two stages of this phase the AE activity is concentrated in the bridge with some activity on the external cracks (panels 4 and 5), this is followed by a period of bridge activity (panel 6) before again the activity spreads outwards, whilst continuing in the bridge (panel 7–8). Failure occurs at about 3.6 hours, after a relatively long delay, and is marked by a large peak in hits and energy with a drop in $^{\Sigma N}/_{\Sigma E}$ and a defocusing of event locations (panel 10).

## *AE Spectral Analysis*

Spectral analysis of the full waveform data (set-up I) was carried out to investigate improved characterisation of event types (e.g., shear versus opening-mode), to help identify phases of failure and look for precursory signals to failure. Figure 18 shows plots of the frequency (log) power spectra for each event in each test from a single transducer. These spectral plots are derived by taking the FFT of the time series relating to each AE event and plotting the power at each frequency as a greyscale value. The power spectra thus derived are plotted for all events together in

▶

Figure 18
Frequency spectra for each full AE waveform, acquired in four of the tests using AE set up I, compared with the loading curves, AE count and fracture development over time. Each figure shows: (i) The curves of force and cumulative AE hits, for the two transducers, versus time; (ii) the fracture development over time from the analysis of digital photographs (where available); (iii) the frequency spectra, derived by FFT of each recorded waveform, plotted as a contour plot with $x$ and $y$ axes of frequency and event number respectively and a greyscale indicating the relative signal (log) power at each frequency.

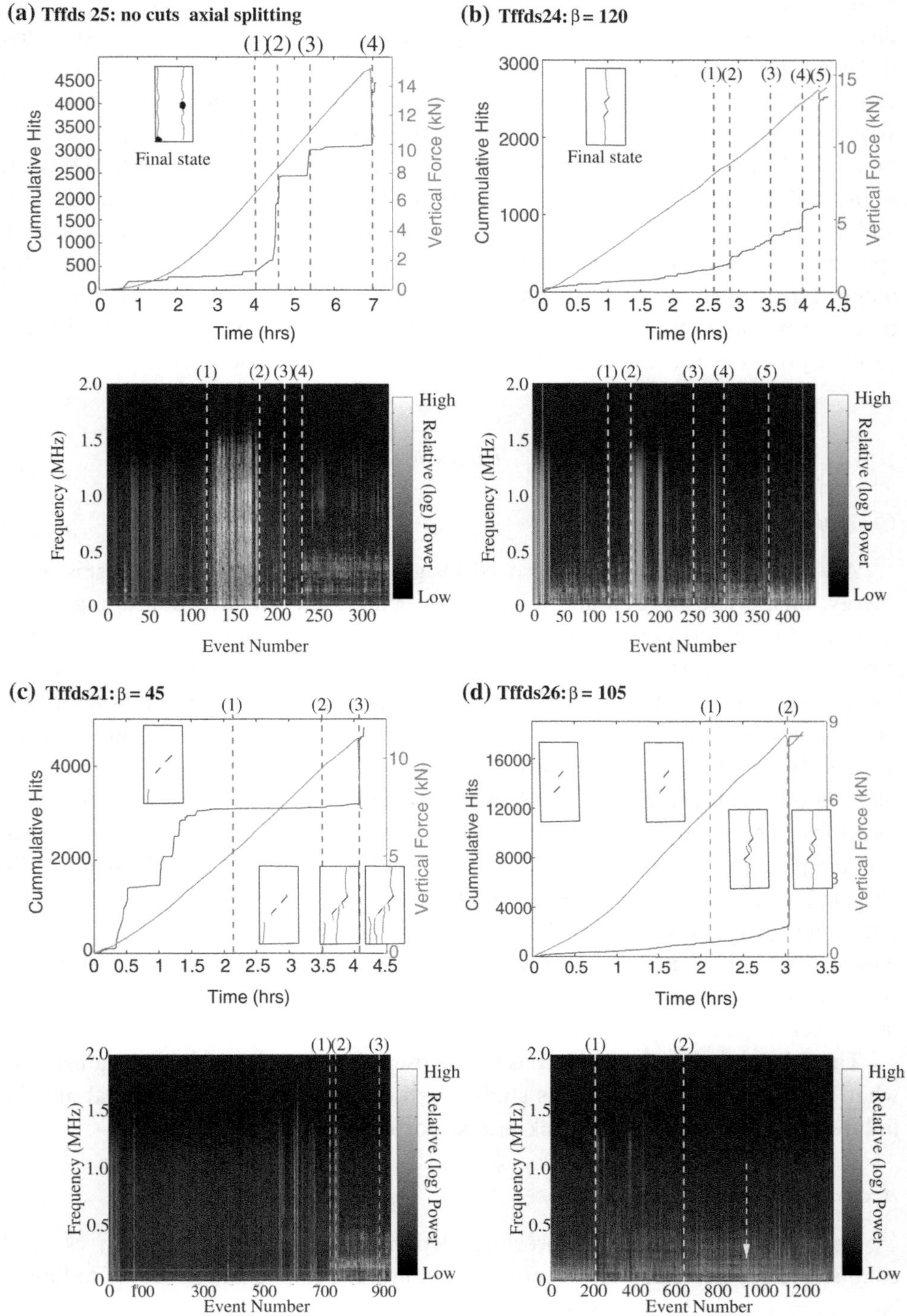
(a) Tffds 25: no cuts axial splitting
(1)(2) (3) (4)
Final state
Cummulative Hits
Vertical Force (kN)
Time (hrs)
Frequency (MHz)
Event Number
High
Relative (log) Power
Low
(b) Tffds24: β = 120
(1)(2) (3) (4)(5)
Final state
Cummulative Hits
Vertical Force (kN)
Time (hrs)
Frequency (MHz)
Event Number
High
Relative (log) Power
Low
(c) Tffds21: β = 45
(1) (2) (3)
Cummulative Hits
Vertical Force (kN)
Time (hrs)
Frequency (MHz)
Event Number
High
Relative (log) Power
Low
(d) Tffds26: β = 105
(1) (2)
Cummulative Hits
Vertical Force (kN)
Time (hrs)
Frequency (MHz)
Event Number
High
Relative (log) Power
Low

a single plot to allow comparison and analysis of the change in energy-frequency distribution through the test. The event spectra are shown versus event number since if the events are plotted with a true time axis it is not possible to discern the different characteristics as the majority events occurred in short duration bursts. Results are shown for 4 tests: *Tffds25* with no pre-existing flaws and failure through opening-mode axial splitting; *Tffds21* ($\beta = 45°$) with a final failure stage by shear (coalescence in the rock bridge); *Tffds24* and *Tffds26* are the two cases with two flaws at $\beta = 120°$ and $\beta = 105°$, respectively. In each case phases can be seen where there are common spectral signatures with clear changes between each phase. Comparison with the mapped fracture development, where available, and the standard AE and load parameters allows the variations in spectral signature to be associated with the rupture processes for these different failure modes, as discussed later.

## *Discussion*

### *General*

AE analysis has provided improved insight into the fracture mechanisms using energy, hits, event location and frequency spectra. Additionally, monitoring of the samples throughout the tests with high resolution digital photographs has allowed direct observation of the failure, which can be linked to the recorded AE and load curves. The combination of AE and photographic monitoring provides improved understanding of the failure processes and coalescence mechanisms in the rock bridge. It is noted that the visual observation of the wing cracks (from the photographs) lags behind that of AE, for example as seen in the patterns of AE event locations. This could relate to the mechanisms of formation of the wing-crack fracture planes in that initially they are planes of microscopic cracks which gradually coalesce, with continued loading, until they are sufficiently open to be seen. This occurs just prior to the initiation of the coalescence in the rock bridge and perhaps indicates that sufficient opening of the wing cracks is necessary before the rock bridge processes can operate. Similarly the rock-bridge coalescence is necessary before total failure can occur.

The following discusses what has been observed, what insight can be gained and what can be concluded from this study, particularly from the AE data, about the failure and coalescence in the rock bridge, for the FGT.

### *Acoustic Emissions*

The results presented in the previous section have shown that AE analyses using counts, energy, etc., can provide more insight into the timing and modes of failure in laboratory tests than can be gained just from load curves, or the photographs, alone. Greater insight has been gained using the ratio ${}^{\Sigma N}/_{\Sigma E}$, event location and spectral analysis. The parameter ${}^{\Sigma N}/_{\Sigma E}$ has good sensitivity to changes in mode of failure for the FGT and is an easily calculated combination of the two key parameters, cumulative hits and energy. Event location has provided a good understanding of the temporal and spatial distribution of events, which allowed the onset of the wing cracks and coalescence in the rock bridge to be identified and distinguished.

Spectral analysis of the AE waveforms has revealed clear differences related to different modes and phases of failure that may be useful in monitoring. For example, in test *Tffds25* three main spectral signatures can be seen. The first is a general very low energy and low frequency type of event that occurs throughout the test but is particularly evident at the beginning and so is likely to be associated with small cracking, grain rearrangement, etc. The second type of event is broad-band and high energy; these begin to occur early in the test but are most significant in the second phase, when the stress-strain response appears to be roughly linear. These events might be associated with opening-mode rupturing or some other mechanism related to the presence of inclusions. A third event type is seen at the end of the test, during the final rupture, and is marked by a moderate bandwidth signature. Broad-band events are also seen in test *Tffds24* ($\beta = 120°$) but in this case the general background event type is low frequency, low energy. The final stage of failure of *Tffds24* involves low frequency events with higher energy, starting at point 5 in Figure 18b, and a single large, broad-band event. Test *Tffds26* ($\beta = 105°$) showed few high energy broad-band events, but has similar background events to test *Tffds24* and the final stage of coalescence in the rock bridge (including the rock-bridge cylinder rotation) seems to have a different spectral character (from point 2 onwards in Fig. 18c); this is moderate bandwidth. The coalescence in the rock bridge for test *Tffds21* ($\beta = 45°$) was through shear failure, which probably explains the markedly different spectral signature in the final phase of this test and seems to have two stages, the first from point 2 and the second marked by the arrow (Fig. 18d).

A general summary classification of the observed events types is:

1. Broad-band events associated with opening-mode ruptures or mechanisms related to the presence of inclusions;
2. low-medium frequency events, possibly associated with tensile fracture propagation;
3. lower frequency events associated with shearing/rotation;
4. background low energy events.

From these observations it appears there is potential for monitoring and characterisation of fracture processes with frequency signatures, but further calibration is required. In particular, tests where spectral analysis, event location and photographic monitoring are all possible concurrently are necessary. *In situ* (i.e., in the cavities below Naples) monitoring and event characterisation using such spectral analysis would probably only be viable near the AE sources due to attenuation of the signal, which will impact primarily the higher frequencies.

It is noted that improved AE analysis could be achieved through a greater number of broad-band transducers, which would allow linking of the event locations and spectral analysis. Also using smaller transducers than were used in set-up II and the event location would provide better definition of event locations by reducing picking and position uncertainty. Finally, focal mechanism analysis, which could not be carried out for these tests, would be insightful to calibrate the event types.

### *Failure Mechanisms*

#### *Interpretation of the Failure in No-flaw and Single-flaw Tests*

The observed non-2-D rupturing of the samples without flaws, as compared to all the tests with two flaws, indicates that the presence of the pre-existing flaws forces the sample failure to be roughly 2-D (in fact only 25% of all the tests showed non-2-D failure, and these were mainly those with no or one pre-existing flaw). Furthermore it is clear that, without any pre-existing flaws, the rupturing is initiated at inclusions. This also confirms the somehow surprising observation that when there are artificial flaws in the system the failure is almost completely controlled by these flaws such that the test results are very repeatable despite the significant inclusion heterogeneity and variations between samples (DE SANCTIS, 2005).

In both the no-flaw and single-flaw cases the failure follows a classical development involving three phases: (1) crack closure; (2) linear stress-strain response and crack initiation with stable crack growth; (3) crack damage and unstable crack growth leading to failure.

When there is a single pre-existing flaw in the sample, the failure mode is through the expected formation and propagation of opening-mode wing cracks from the flaw tips. This is a similar process to that initially observed in the samples with two flaws until the two internal wing cracks start to interact, at which point the mechanisms of coalescence in the rock bridge are key to the overall sample failure.

#### *Interpretation of the Failure in Tests with Two Pre-existing Faws*

Comparison of the experimentally identified failure phases, for the samples with no or one flaw and those with two flaws, shows that the first two phases are quite similar whilst the third one is different due to the rock bridge mechanisms. In the cases with two flaws, the third, non-linear phase is split into two stages. In fact, the failure of the sample occurs only after both the unstable propagation of external wing cracks and coalescence of the internal cracks in the bridge. The duration of the unstable propagation of the external cracks is short when there are two pre-existing flaws. The timing of the coalescence phase is characterised by the time between two significant jumps in AE and depends on the flaw geometry. It is only using the AE signatures that it is possible to determine these time ranges, which are summarised as follows:

**Phase 1:** Closure of existing (natural) cracks and flaws, which normally finishes abruptly with a decrease in ${}^{\Sigma N}/_{\Sigma E}$ and is followed by a period of little AE activity or deformation.

**Phase 2:** Initiation of wing cracks, at both the internal and external flaw tips, and propagation of external wing cracks. The start of this phase, after a short period of little activity, is marked by increasing AE hits, energy and, most clearly, ${}^{\Sigma N}/_{\Sigma E}$. A significant increase in AE marks the end of the phase. By the end of the phase some propagation of the external wing cracks can be seen parallel to the loading direction and away from the flaws.

**Phase 3:** Development of the interaction cracks and coalescence in the rock bridge leading to final crack propagation and rupture. A significant jump in AE marks the onset of the coalescence of the internal/interaction cracks in the rock bridge. Coalescence in the rock bridge leads to the final jump in AE, which is the largest and marks the sample failure due to the final propagation of all the cracks.

The start of phases 2 and 3 are marked by positive increases/jumps in ${}^{\Sigma N}/_{\Sigma E}$, indicating an increase in the number of events and/or a decrease in the energy of each event. At the final failure there is generally a large negative jump in ${}^{\Sigma N}/_{\Sigma E}$ indicating a single or a few large events. In terms of the timing and duration of each phase it is seen that phases 1 and 2 have little dependence on the flaw configuration (the duration of phase 1 is probably related to the inclusions). However, phase 3 seems to depend directly on the flaw configuration as it has a shorter duration for $\beta = 120°$ than for $\beta = 105°$. These differences are probably due to the different modes of coalescence in the rock bridge. For $\beta = 105°$ the total failure is preceded by bridge rotation, which appears to take longer than the simpler mode of failure for $\beta = 120°$. Note also that the peak stress is also generally lower for $\beta = 105°$ than $\beta = 120°$.

## *Conclusions*

This work has shown that monitoring of rock-sample failure (in this case for the Neapolitan fine-grained tuff) with a combination of high resolution photographs and AE, even with just standard AE attributes (such as energy and hits), can provide improved insight into failure processes, fracture mechanisms and coalescence in rock bridges. Additional AE attributes, such as the ratio ${}^{\Sigma N}/_{\Sigma E}$, event location and spectral analysis (of the full AE waveform data), have been shown to be even more insightful. The former seems to provide clear indications of changes in deformation phase for these tests. AE event location has provided better information on timing and positions of failure. Spectral analysis of waveforms has revealed different event characteristics that may be useful in monitoring. The digital photographs provided clear and direct evidence of fracturing and fracture pattern development, although it is noted that the visual verification, with the photographs, lags behind the identification of fracture planes with the AE data. This lag suggests that *different scales* of fracture formation are visible by the *different methods*. Combining event location and spectral analysis with the photographic observations will, in the future, provide even better characterisation of fracture initiation, propagation and coalescence plus enable improved characterisation of AE event types for use as an *in situ* monitoring system.

Based on the observations made, it can be concluded that, for the Neapolitan fine-grained tuff, failure, in the presence of pre-existing flaws is highly dependent on the juxtaposition of the flaws, however the different modes of failure can be well characterised in the laboratory. As such, a general description of the fracture development and coalescence in the rock bridge has been identified with three phases similar to that in a classical description. However, the final phase leading to rupture, when there are two pre-existing flaws, involves an additional mechanism for the coalescence of wing cracks across the bridge. The timing and mode of this final coalescence is strongly dependent on the pre-existing flaw geometry, e.g., the duration is shorter for $\beta = 120°$ than for $\beta = 105°$ and for $\beta = 105°$ the total failure is preceded by a bridge rotation. Few precursory warning signals to failure have been identified, even in the AE analysis. *In situ* any precursory signals are most likely to be from AE signatures, in particular their location, frequency character and the ${}^{\Sigma N}/_{\Sigma E}$ parameter, detected with AE sensors local to high risk fracture zones. Such data might allow assessment of what stage/type of failure is occurring and so the state of coalescence of the fracturing might be inferred. Even though the observations provided in this work show that the coalescence mechanisms may lead very rapidly to failure with only slight prior warning from the AE monitoring, *in situ* collapses will occur as a result of many such rock-bridge failures and under different stress conditions. Therefore monitoring with AE within the cavities below Naples may allow tracking of rock-bridge failures and assessment of the overall likelihood of block detachment and large failures.

## *Acknowledgments*

Thank you to Pascal Charrier, for his superb assistance in the lab, and to Marc Boulon (Laboratoire 3S) and André Vervoort (KU Leuven) for the use of their AE transducers. FdS acknowledges the invaluable insight into the mechanical behaviour of the tuff of Aldo Evangelista (Università di Napoli Federico II). The Commissariat à l'Energie Atomique (CEA) in Grenoble is acknowledged for the collaboration and use of their wire saw. SH was supported by a Marie Curie Intra-European Individual Fellowship within the 6th European Community Framework Programme (project iGEM). We also are grateful for the detailed and very helpful feedback from Sergei Stanchits and another, anonymous, reviewer.

## *Appendix*

Figure 19 shows b values for all transducers in test *Tffds34* (AE set-up II) throughout the duration of the test, calculated using a window length of 500 events and a step of 100 events between windows. The mean time of each analysis window has been used to plot the values, as by other authors previously, although it is noted that the results depend on the windows chosen see later comments. As discussed in the main text, the b values calculated are higher than often observed for other rocks.

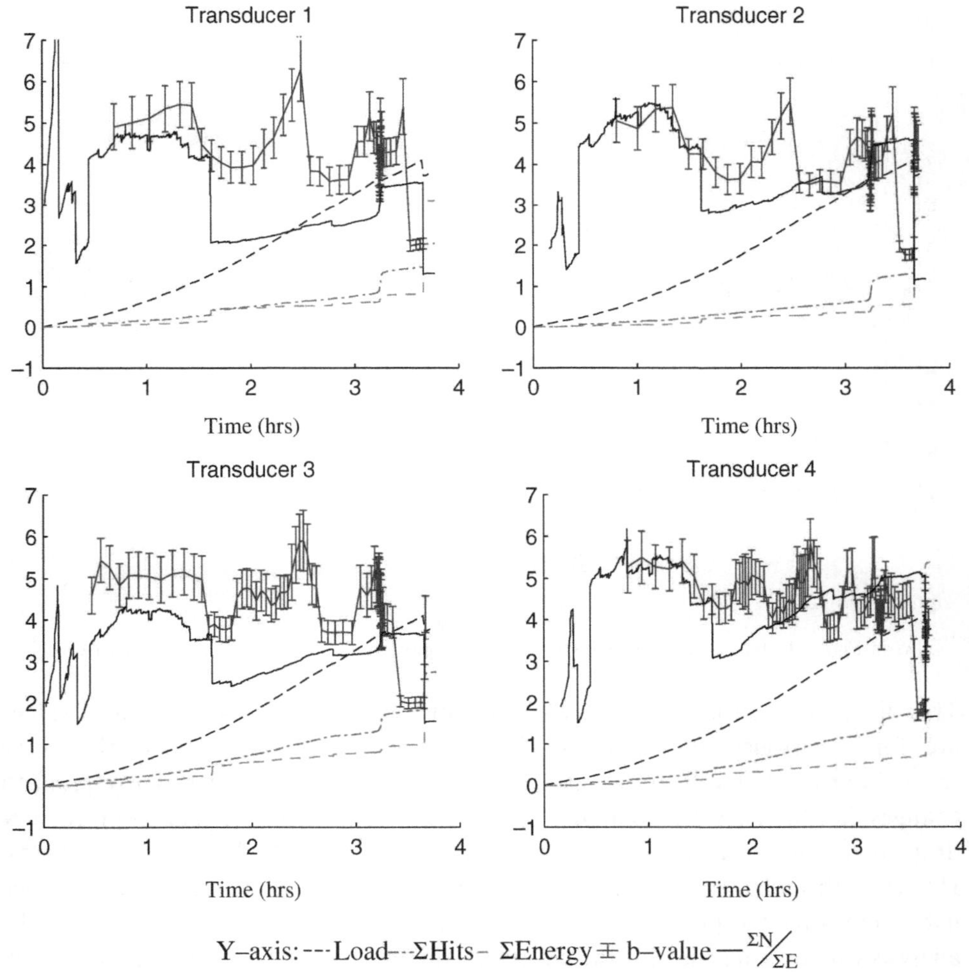


Figure 19

Comparison of the b-values for all transducers in test *Tffds34* (AE set-up II) using the mean time of each window to plot the values, a window length of 500 events and a step of 100 events between windows. The error bars relate to the standard deviation from the least-squares solution of the line fitting. The vertical force, $\Sigma N$, $\Sigma E$, and ${}^{\Sigma N}/_{\Sigma E}$ are also given but as scaled (relative) values.

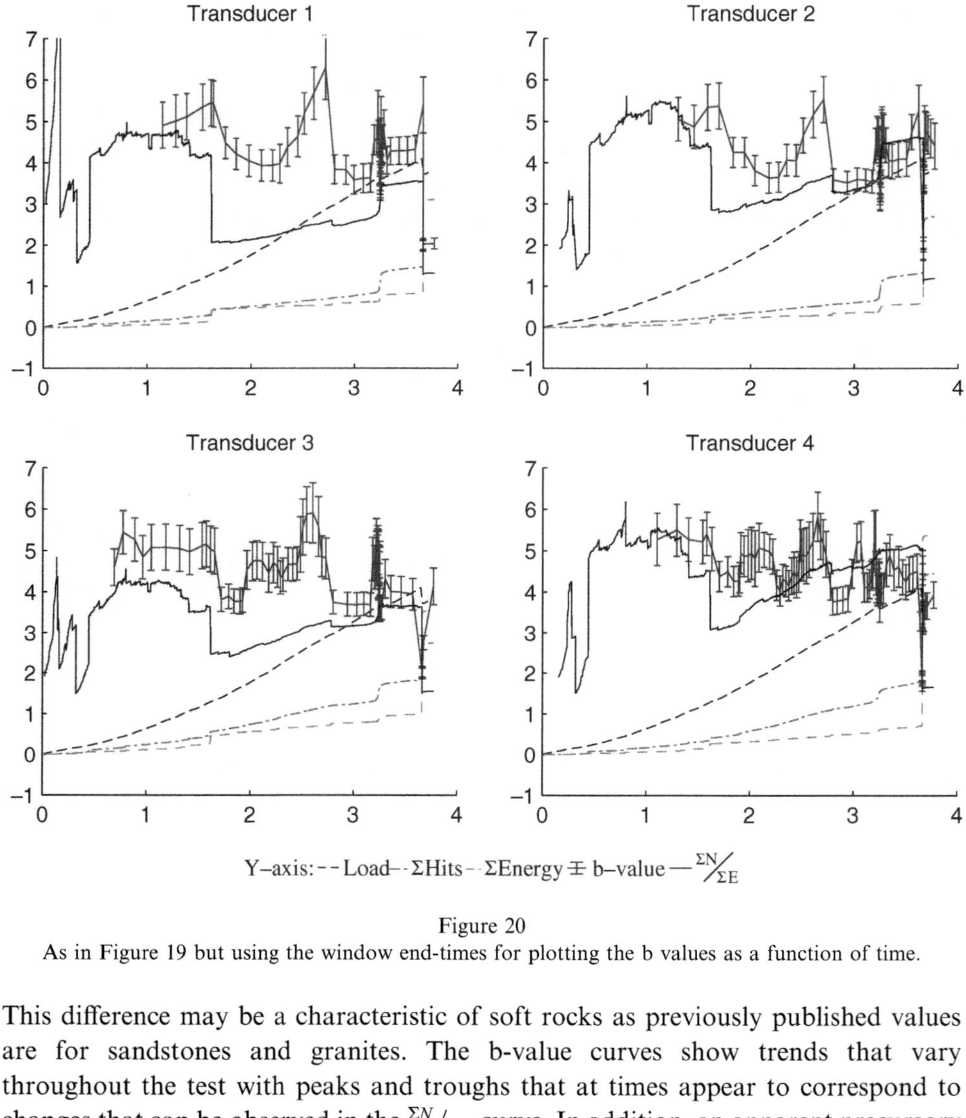


Figure 20
As in Figure 19 but using the window end-times for plotting the b values as a function of time.

This difference may be a characteristic of soft rocks as previously published values are for sandstones and granites. The b-value curves show trends that vary throughout the test with peaks and troughs that at times appear to correspond to changes that can be observed in the $^{\Sigma N}/_{\Sigma E}$ curve. In addition, an apparent precursory drop in b value can be seen, which is not seen in the $\Sigma N$, $\Sigma E$ or $^{\Sigma N}/_{\Sigma E}$ curves. However, this seems to in part arise from the acausal nature of the approach, since data from after the plotting time are used in the analysis. If the end-times of the analysis windows are used for the plotting, a different picture is seen (Fig. 20), which resembles more that of $^{\Sigma N}/_{\Sigma E}$, and in particular the precursory drop is not seen, the drop actually seems to occur, as with $^{\Sigma N}/_{\Sigma E}$, at the point of failure. Thus similar insight can be gained from $^{\Sigma N}/_{\Sigma E}$ as for the b value and, certainly for the case of the FGT, the different phases of deformation are more clearly defined using $^{\Sigma N}/_{\Sigma E}$. Importantly the $^{\Sigma N}/_{\Sigma E}$ is also simple to calculate and involves no line fitting.

## References

AMITRANO, D. (2003), *Brittle-ductile transition and associated seismicity: Experimental and numerical studies and relationship with the b value,* J. Geophys. Res. *108*, 2044–2059.

ASSOCIAZIONE GEOTECNICA ITALIANA. *Raccomandazioni sulla programmazione ed esecuzione delle indagini geotecniche* (Associazione Geotecnica Italiana, Roma, 1977).

AVERSA, S. and EVANGELISTA, A. (1998), *Mechanical behaviour of a volcanic tuff: Yield, strength and destructuration effect,* Rock Mech. Rock Eng. *31*, 25–42.

AVERSA, S., EVANGELISTA, A., and RAMONDINI, M. (1991), *Snervamento e resistenza a rottura di un tufo a grana fine,* Atti del II Convegno dei Ricercatori del G.N.C.S.I.G. del C.N.R., Ravello *1*, 3–22.

BAUD, P., KLEIN, E., and WONG, T-F. (2004), *Compaction localization in porous sandstones: spatial evolution of damage and acoustic emission activity,* J. Struct. Geol. *26*, 603–624.

BIENIAWSKI, Z.T. (1967), *Mechanism of brittle fracture of rock, parts I, II and III,* Int J Rock Mech. Min. Sci. Geomech. Abstr. *4*, 395–430.

BOBET, A. (2000), *The initiation of secondary cracks in compression,* Engin. Fracture Mech. *66*, 187–219.

BOBET, A. and EINSTEIN, H.H. (1998), *Fracture coalescence in rock-type materials under uniaxial and biaxial compression,* Int. J. Rock Mech. Min. Sci. *35*, 863–888.

CARVALHO, F.C.S. and LABUZ, J.F. (2002), *Moment tensors of acoustic emissions in shear faulting under plane-strain compression,* Tectonophysics *356*, 199–211.

COLLINS, D.S., PETTITT, W.S., and YOUNG, R.P. (2002), *High-resolution mechanics of a microearthquake sequence,* Pure Appl. Geophys. *159*, 197–219.

DE SANCTIS, F. (2005), *Studio dei meccanismi di rottura fragile nel tufo Napoletano,* Ph.D. Thesis, Università di Napoli Federico II, Napoli, Italy.

DESRUES, J. (1984), *La localisation de la déformation dans les milieux granulaires,* Ph.D. Thesis, Université de Grenoble, Grenoble, France.

DESRUES, J. and VIGGIANI, G. (2004), *Strain localization in sand: an overview of the experimental results obtained in Grenoble using stereophotogrammetry,* Int. J. Num. Anal. Meth. Geomech. *28*, 279–321.

EBERHARDT, E., STEAD, D., STIMPSON, B., and READ, R.S. (1998), *Identifying crack initiation and propagation thresholds in brittle rock,* Can. Geotech. J. *35*, 222–233.

GEOLOGICAL SOCIETY ENGINEERING GROUP WORKING PARTY (1970), *The logging of rock cores for engineering purposes,* Q. J. Engng. Geol. *3*, 1–24.

GLEA (2002), *Acoustic Emission Reference Guide,* Grupo Latinoamericano de emision acustica (GLEA) (http://www.cnea.gov.ar/cac/endye/glea/mateo/guide.htm).

HATTON, C.G., MAIN, I.G., and MEREDITH, P.G. (1993), *A comparison of seismic and structural measurements of scaling exponents during tensile subcritical crack-growth,* J. Struct. Geol. *15*, 1485–1495.

HAZZARD, J.F. and YOUNG, R.P. (2002), *3D numerical modeling of acoustic emissions.* In Proc. 5th Int. Workshop Application of Geophysics in Rock Engineering, July 7 2002, Toronto, Canada, pp. 40–47.

INGRAFFEA, A.R. and HEUZE, F.E. (1980), *Finite element models for rock fracture mechanics,* Int. J. Num. Anal. Methods Geomech. *4*, 25–43.

JIEFAN, H., GANGLIN, C., YONGHONG, Z., and REN, W. (1990), *An experimental study of the strain field development prior to failure of a marble plate under compression,* Tectonophysics *175*, 269–284.

LABUZ, J.F., CATTANEO, S., and CHEN, L-H. (2001), *Acoustic emission at failure in quasi-brittle materials,* Construction and Building Materials *15*, 225–233.

LEI, X., KUSUNOSE, K., RAO, M.V.M.S, NISHIZAWA, O., and SATOH, T. (2000), *Quasi-static fault growth and cracking in homogeneous brittle rock under triaxial compression using acoustic emission monitoring,* J. Geophys. Res. *105*, 6127–6139.

LEI, X., MASUDA, K., NISHIZAWA, O., JOUNIAUX, L., LIU, L., MA, W., SATOH, T., and KUSUNOSE, K. (2004), *Detailed analysis of acoustic emission activity during catastrophic failure of faults in rock,* J. Struct. Geol. *26*, 247–258.

LIAKOPOULOU-MORRIS, F., MAIN, I.G., CRAWFORD, B.R., and SMART, B.G.D. (1994), *Microseismic properties of a homogeneous sandstone during fault nucleation and frictional sliding,* Geophys. J. Int. *119*, 219–230.

LOCKNER, D.A. (1993), *The role of acoustic emissions in the study of rock fracture,* Int. J. Rock Mech. Min. Sci. Geomech. Abstr. *30*, 883–889.

LOCKNER, D.A. and BYERLEE, J.D. (1977), *Hydrofracture in Weber sandstone at high confining pressure and differential stress,* J. Geophys. Res., *82*, 2018–2026.

LOCKNER, D.A., BYERLEE, J.D., KUKSENKO, V., and PONOMAREV, A., *Observations of quasistatic fault growth from acoustic emissions,* In Fault Mechanics and Transport Properties of Rocks (eds. Evans, B. and Wong, T-F.) (Academic Press, London, 1992) pp. 3–31.

LOCKNER, D.A., BYERLEE, J.D., KUKSENKO, V., PONOMAREV, A., and SIDORIN, A. (1991), *Quasi-static fault growth and shear fracture energy in granite,* Nature *350*, 39–42.

MAIN, I.G., MEREDITH, P.G., and JONES, C. (1989), *A reinterpretation of the precursory seismic b value anomaly from fracture mechanics,* Geophys. J. Int. *96*, 131–138.

MUGHIEDA, O. and ALZOUBI, K.A. (2004), *Fracture mechanisms of offset rock joints — A laboratory investigation,* Geotech. Geol. Eng. *22*, 545–562.

PETIT, J. and BARQUINS, M. (1988), *Can natural faults propagate under mode II conditions?* Tectonics *7*, 1243–1256.

PHYSICAL ACOUSTICS CORPORATION (2001), MISTRAS2001 *Users Manual.*

REYES, O. and EINSTEIN, H.H. (1991), *Failure mechanism of fractured rock - a fracture coalescence model,* In Proceedings 7th International Congress of Rock Mechanics, pp. 333–340.

SCAVIA, C. and CASTELLI, M. (1998), *Studio della propagazione per trazione indotta di sistemi di fratture in roccia,* Rivista Italiana di Geotecnica, anno XXXII, 48–62.

SCHOLZ, C.H. (1968), *Microfracturing and the inelastic deformation of rock in compression,* J. Geophys. Res. *73*, 1417–1432.

SHEN, B. and STEPHANSSON, O. (1993), *Large-scale permeability tensor of rocks from induced microseismicity,* Int. J. Rock Mech. Min. Sci. Geomech. Abstr. 30, 861–867.

SHEN, B., STEPHANSSON, O., EINSTEIN, H.H., and GHAHREMAN, B. (1995), *Large-scale permeability tensor of rocks from induced micro-seismicity,* J. Geophys. Res. *100*, 5975–5990.

SHIOTANI, T., YUYAMA, S., LI, Z.W., and OTHSU, M., *Quantitative evaluation of fracture process in concrete by the use of improved b value,* In *5th Int. Symposium Non-Destructive Testing in Civil Engineering* (ed. Uohoto) (T. Elsevier Science, Amsterdam, 2000) pp. 293–302.

STANCHITS, G. and DRESEN, G. (2003), *Separation of tensile and shear cracks based on acoustic emission analysis of rock fracture,* In *Int. Symp.* (*NDT-CE 2003*), *Non-destructive Testing in Civil Engineering 2003,* 107 pp.

WONG, R.H.C. and CHAU, K.T. (1998), *Crack coalescence in a rock-like material containing two cracks,* Int. J. Rock Mech., Mi. Sci. *35*, 147–164.

ZANG, A., WAGNER, F.C., and DRESEN, G. (1996), *Acoustic emission, microstructure, and damage model of dry and wet sandstone stressed to failure,* J. Geophys. Res. *101*, 17507–17521.

ZANG, A., WAGNER, F.C., STANCHITS, S., DRESEN, G., ANDRESEN, R., and HAIDEKKER, M.A. (1998), *Source analysis of acoustic emissions in Aue granite cores under symmetric and asymmetric compressive loads,* Geophys. J. Int. *135*, 1113–1130.

(Received June 22, 2005, revised January 17, 2006, accepted January 18, 2006)

Pure appl. geophys. 163 (2006) 2205–2221
0033–4553/06/102205–17
DOI 10.1007/s00024-006-0118-y



Pure and Applied Geophysics

# Understanding the Seismic Velocity Structure of Campi Flegrei Caldera (Italy): From the Laboratory to the Field Scale

S. Vinciguerra,[1] C. Trovato,[2] P.G. Meredith,[3] P.M. Benson,[3]
C. Troise,[2] and G. De Natale[2]

*Abstract*—We report laboratory measurements of P- and S-wave velocities on samples of tuff from Campi Flegrei (Italy), and a new tomographic velocity map of the Campi Flegrei caldera. Laboratory measurements were made in a hydrostatic pressure vessel during both increasing and decreasing effective pressure cycles. Selected samples were also thermally stressed at temperatures up to 600°C to induce thermal crack damage. Acoustic emission output was recorded throughout each thermal stressing experiment, and velocities were measured after thermal stressing. Laboratory P- and S-wave velocities are initially low for the tuff, which has an initial porosity of ~45%, but both increase by between 25 and 50% over the effective pressure range of 5 to 80 MPa, corresponding to a decrease of porosity of ~70%. Marked velocity hysteresis, due to inelastic damage processes, is also observed in samples subjected to a pressurization-depressurization cycle. Tomographic seismic velocity distributions obtained from field recordings are in general agreement with the laboratory measurements. Integration of the laboratory ultrasonic and seismic tomography data indicates that the tuffs of the Campi Flegrei caldera can be water or gas saturated, and shows that inelastic pore collapse and cracking produced by mechanical and thermal stress can significantly change the velocity properties of Campi Flegrei tuffs at depth. These changes need to be taken into account in accurately interpreting the crustal structure from tomographic data.

**Key words:** Laboratory P- and S-wave velocities, tomographic velocity map, mechanical and thermal stresses.

## *1. Introduction*

The Campi Flegrei caldera, located a few km west of the City of Naples, Southern Italy (Fig. 1), is currently one of the most important areas worldwide for detailed volcanological studies. This is because significant uplift (rates up to 1 m/year) and slow secular subsidence, recognized from marine encrustations on ancient Roman

[1] HP-HT Laboratory Istituto Nazionale di Geofisica e Vulcanologia, sezione di Roma 1, Via di Vigna Murata 605, 00143, Rome, Italy. E-mail: vinciguerra@ingv.it
[2] Istituto Nazionale di Geofisica e Vulcanologia, sezione di Napoli, Via Diocleziano 328, 80124, Naples, Italy.
[3] Department of Earth Sciences, University College London, Gower Street, London WC1E6BT, London, United Kingdom.

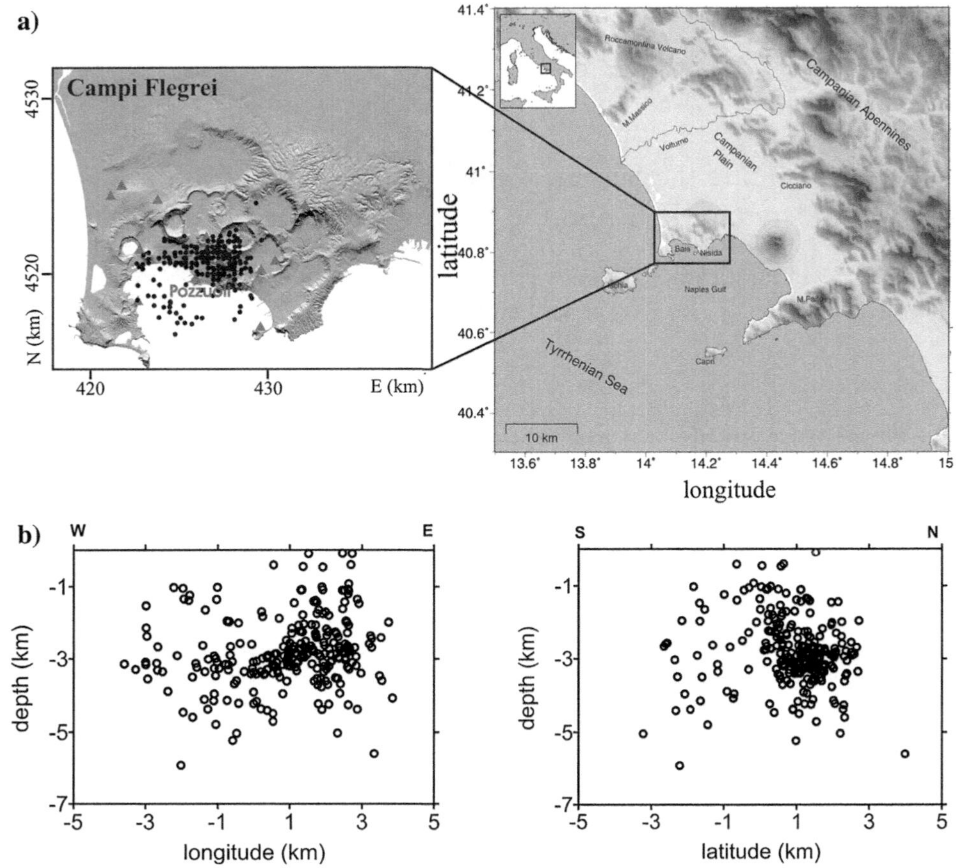


Figure 1
(a) Map of the Campanian volcanic area. The inset shows the locations of the permanent recording stations operating in the Campi Flegrei area; (b) E-W and N-S depth sections of locations for the 331 Campi Flegrei earthquakes used in this study, determined by the Bayloc method (PRESTI *et al.*, 2005).

remains, indicate the possibility of renewed volcanic activity that may pose a serious threat to the densely-inhabited Neapolitan area.

Campi Flegrei is a nested caldera formed by a sequence of collapse episodes. The formation of the largest caldera, with a radius of about 6 km, is related to the collapses that followed the eruption of the Campanian Ignimbrite around 35,000 years ago (ROSI and SBRANA, 1987). The innermost collapse occurred about 12,000 years ago, and is associated with the major eruption of the Neapolitan yellow tuff. The activity at the Campi Flegrei caldera has been characterized by many explosive eruptions from different vents. The last eruption occurred in 1538 AD after several years of rapid ground uplift, and resulted in a high level of damage to buildings and loss of life. More recently, two very large uplift episodes, apparently not followed by eruptions, occurred between 1970 and 1984. These were followed by a period of rapid subsidence (around 5 to 8 cm/year) that is still in progress.

The scientific community has recently increased its efforts in seeking to understand the eruptive hazard associated with these uplift-subsidence events, both by detailed analysis of ground deformation (BIANCHI *et al.*, 1987; DVORAK and BERRINO, 1991; BONAFEDE *et al.*, 1986; DE NATALE *et al.*, 1991, 1997; DE NATALE and PINGUE, 1993; BONAFEDE and MAZZANTI, 1998) and seismicity (TROISE *et al.*, 1997; 2003), and by modelling of the overall physics and mechanics of the process (GAETA *et al.*, 1998, 2003; DE NATALE *et al.*, 2001). One of the key issues related to the recent activity lies in determining the role played by water circulation and thermal-fluid perturbations (BONAFEDE and MAZZANTI, 1998; GAETA *et al.*, 1998; 2003; DE NATALE *et al.*, 2001, TROISE *et al.*, 2001; ASTER and MEYER, 1988). In order to address this, and to be able to model the processes acting within the caldera, detailed knowledge of the crustal structure and the properties of the relevant rocks is required.

In an attempt to reconstruct the crustal structure, ASTER and MEYER (1998), presented a three-dimensional velocity model for the central part of the caldera. From their data, they inferred that the rocks were likely to be water saturated. However, their interpretation should be treated with caution, since the resolution was limited by the relatively small number of earthquakes and recording stations. A recent tomographic study performed with the 1982–1984 microearthquakes subdata set confirmed the presence of anomalous velocity values, which are petrophysically interpreted as overpressured gas-bearing formations (VANORIO *et al.*, 2005).

ZAMORA *et al.* (1994) provided the first laboratory measurements of ultrasonic wave velocities on rocks from Campi Flegrei. Compressional (Vp) and shear (Vs) wave velocities were measured at ambient pressure and temperature on core samples from four boreholes, and from a maximum depth of 3 km. Again, they suggested that velocities in water-saturated samples were more representative of *in situ* rock conditions than those in dry samples. Differences between their laboratory and available *in situ* velocities were interpreted as being due to velocity dispersion between ultrasonic and sonic frequencies and by the elevated pressure and temperature conditions existing at depth. Unfortunately, the location of the wells (at the boundaries of the caldera) and their limited areal distribution do not allow their data to be used for improving the velocity models necessary for relocation purposes (PUJOL, 2000), despite the caldera symmetry (ZAMORA *et al.*, 2001). More importantly, velocity variations with depth, which are crucial for relocation techniques, could not be analyzed by ZAMORA *et al.* (1994), because all their measurements were made only at ambient pressure.

More recently, VANORIO *et al.* (2002) presented laboratory measurements of both P- and S-wave velocities for a number of volcanic rocks from Italy, including Neapolitan yellow tuff, the main lithology in the Campi Flegrei caldera. Their measurements were made at confining pressures up to 60 MPa and effective pressures up to 45 MPa. However, their results are discussed specifically in terms of the activation of different microstructural deformation mechanisms under dry and saturated conditions, and are not used to address the issue of crustal structure.

Here we present a systematic set of laboratory measurements of Vp and Vs made on samples of the Campi Flegrei tuff. Measurements were made over a range of effective pressures up to 80 MPa. We further investigated the effect of temperature on the physical properties of Campi Flegrei tuff by thermal stressing since this can also modify the microstructure by inducing thermal crack damage.

Our overall aim was to simulate the role of increasing depth on the physical properties of the tuff, which is crucial for the reconstruction of crustal structure from tomographic data. Hence, the laboratory data are interpreted jointly with *in situ* P-wave velocities calculated from a new, local earthquake tomography model that incorporates the largest data set available (331 earthquakes) for the area. One of the key questions we address, from the comparison of laboratory and *in situ* measurements, is whether the tuffs are dry or water saturated, as previously hypothesized by Zamora *et al.* (1994) on the basis of laboratory measurements and by the tomographic studies of Aster and Meyer (1988).

## 2. *Laboratory Scale Experimental Investigation*

### *2.1. Methodology*

The material investigated was the Campi Flegrei tuff (12 ka), which is the most abundant and widely distributed lithology in the caldera (Rosi and Sbrana, 1987). It is a strongly heterogeneous pyroclastic flow material, which includes cavities, pumice and crystals of sanidine, pyroxene and biotite (Vanorio *et al.*, 2002). It has an initial porosity of around 45%. In order to take into account the variability due to the heterogeneity of the Campi Flegrei tuff, three blocks of material were collected from depths of a few tens of meters from a quarry located near the village of Pozzuoli. Laboratory measurements were subsequently made on core samples 38 mm in diameter by 40 mm long, taken from these blocks.

Ultrasonic wave-velocity measurements were made by the pulse-transmission technique inside a fluid-medium hydrostatic pressure vessel. A diagram of the internal arrangement is given as Fig. 2. A 900 volt pulser was used to excite two 1 MHz piezo-electric transmitting transducers at one end of the sample, one each for compressional and shear modes. Signals were received by two identical transducers located at the opposite end of the sample. Error in the first arrival time is estimated as approximately 1% for P waves and about 3% for S waves (Benson *et al.*, 2005). Measurements were carried out on both dry and wet samples at effective pressures up to 80 MPa. Here we define the effective pressure as simply the difference between the confining pressure and the pore fluid pressure. Dry measurements were made on

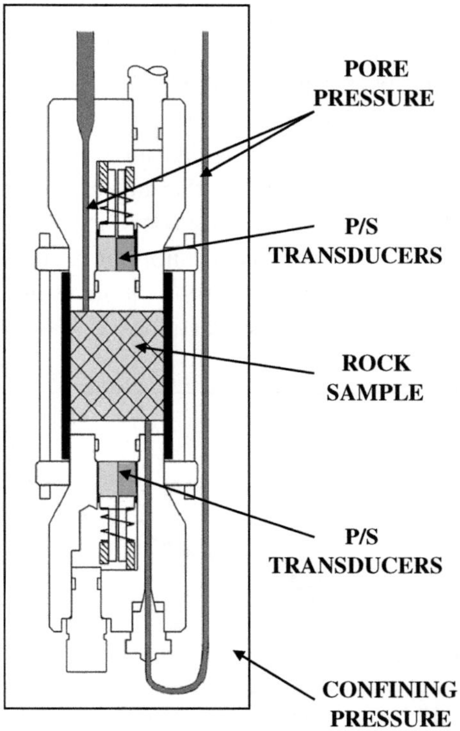


Figure 2
Schematic diagram of the measurement assembly for velocity and pore fluid measurements within the hydrostatic pressure vessel (BENSON *et al.*, 2005).

samples that had been dried in an oven at 80°C for a minimum of 24 hours. Previous experience suggests that little if any thermal crack damage occurs at this temperature. Wet measurements were made on samples that had been vacuum-saturated with distilled water for a minimum of 24 hours. Since the tuff is mechanically very weak, a low pore fluid pressure of 1.5 MPa was used throughout in order to prevent damage to the microstructure. Effective pressure was increased in steps of 5 MPa up to 40 MPa and steps of 10 MPa for all higher pressures. The change in sample porosity was measured from the volume of water expelled during each incremental increase in effective pressure. This decrease in pore volume is attributed to the closure of cracks and pores with increase in pressure (BENSON *et al.*, 2005).

### *2.2. Thermal Treatment*

Selected samples were thermally treated by heating either to 300° or 600°C at a rate of 1°C/min in a tube furnace at room pressure. In order to monitor the amount of induced thermal damage, the output of acoustic emissions (AEs) was recorded throughout each thermal stressing experiment (VINCIGUERRA *et al.*, 2005).

Figure 3 shows that relatively continuous AE activity is observed up to about 200°C. This is followed by a general decrease in the AE rate, but with bursts of activity around 400°C and from 540° to 600°C. This suggests the action of different deformation mechanisms. The lower temperatures activity is likely to be due to thermal expansion mismatch between different mineral phases, while dehydration of the zeolite phases present and associated thermal cracking are considered to be responsible for the higher temperature bursts (DE GENNARO *et al.*, 1999). However, study of the time evolution of AE activity was not our primary goal in this study. We merely use the thermal stressing as a method of investigating thermally-induced damage in our samples, and its effect on elastic wave velocities.

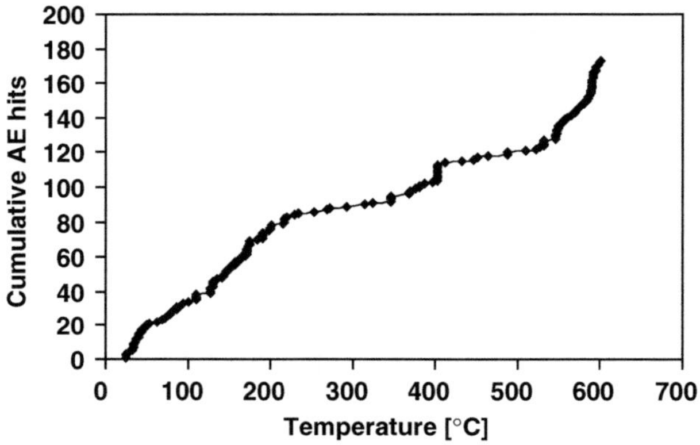


Figure 3
Cumulative number of AE hits, as functions of temperature for heat treatment of a tuff sample to 600°C.

## *2.3. Results*

Simultaneous measurements of P- and S-wave ultrasonic velocities on both dry and saturated samples have been made at effective pressures from 5 to 80 MPa during both pressurization and depressurization cycles. Measurements were made on the following samples types: (a) Non-thermally treated, (b) thermally treated to 300°C and (c) thermally treated to 600°C. The change in length of samples resulting from the application of hydrostatic pressure has been calibrated with respect to the changes in porosity, and times-of-flight corrected for this effect, with a reduction in sample length of about 10% at the maximum effective pressure of 80 MPa and a recovery of about (10%). All velocities reported here have been corrected for the effect of sample shortening.

Figures 4 and 5 show, respectively, P- and S-wave velocities measured on dry samples as a function of confining pressure. At the lowest pressures it was often difficult to pick the first arrival due to its low amplitude. The process became progressively easier as the amplitude increased at higher pressures. Initial velocities are very low, as would be expected for a material with an initial porosity of around

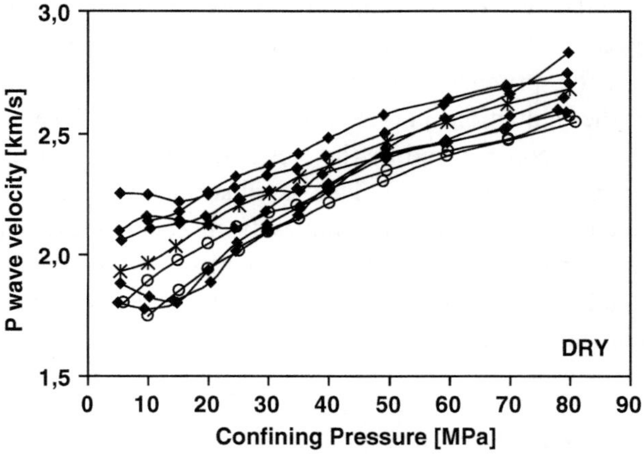


Figure 4
Plot of P- wave velocities as a function of confining pressure for samples tested dry. Solid symbols denote non-heat-treated samples, asterisks denote samples heat-treated to 300°C, and open symbols denote samples heat-treated to 600°C.

45%. However, all the velocities increase significantly with increasing confining pressure; by between 25 and 50% over the pressure range 5 to 80 MPa. This is true for both P and S waves, for samples from all three blocks and for both heat-treated and non-heat-treated samples. Samples that had been heat-treated to 600°C (open symbols on Figs. 4 and 5) had velocities lower than those of non-heat-treated samples. This is more evident for S waves (Fig. 5) than for P waves (Fig. 4). By contrast, samples that had been heat-treated to only 300°C (asterisks on Figs. 4

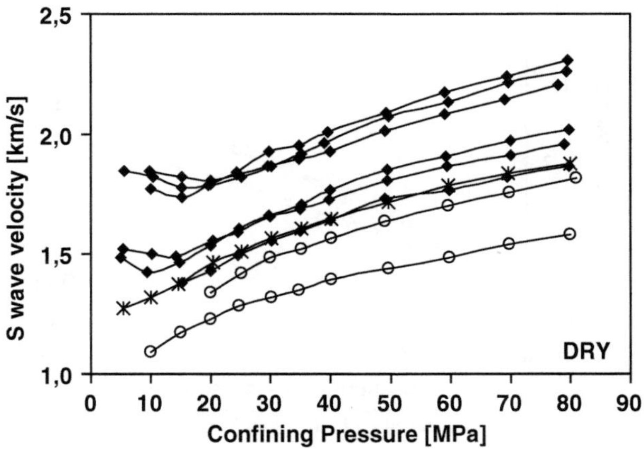


Figure 5
Plot of S- wave velocities as a function of confining pressure for samples tested dry. Solid symbols denote non-heat-treated samples, asterisks denote samples heat-treated to 300°C, and open symbols denote samples heat-treated to 600°C.

and 5) show similar velocities to non-heat-treated samples. These observations support the suggestion that dehydration of the zeolite phases and associated cracking above 400°C is the primary deformation mechanism producing both acoustic emissions and the reduction in velocity. The pattern of velocity change with increasing pressure is complicated by the fact that different changes are observed for the three different sample blocks. This is illustrated in Fig. 6, where averaged normalized velocities are given for samples from each block. These observations support the idea that the compaction mechanism responsible for the velocity increases is strongly influenced by the material heterogeneity and especially the large inclusions within the rock matrix.

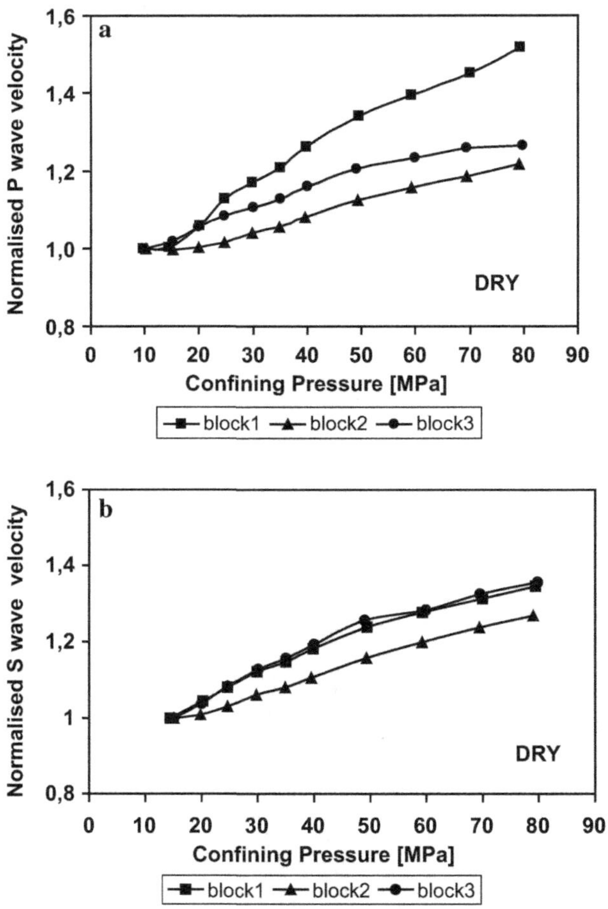


Figure 6
Plots of averaged (a) P- and (b) S-wave velocities for the three different sample blocks as a function of confining pressure. Different symbols are used to distinguish the three blocks. All data are for dry measurements from Figures 4 and 5.

Figure 7 shows an example of the significant hysteresis in velocity that is observed when samples are first pressurized to 80 MPa and then depressurized. Both Vp and Vs are about 10% higher after depressurization than before initial pressurization. This implies that at least part of the deformation responsible for the sample compaction is inelastic and irreversible.

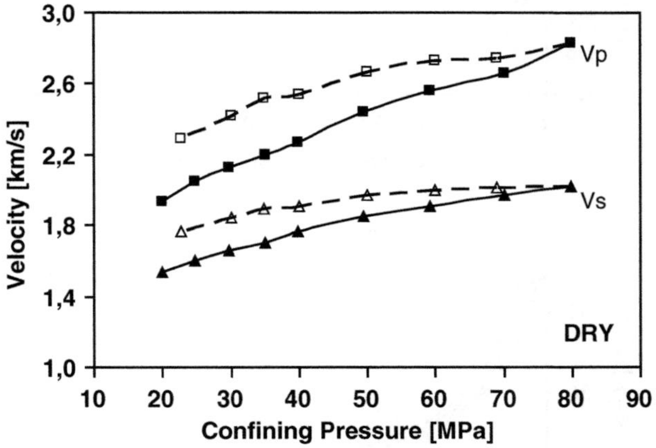


Figure 7
Example of velocity hysteresis for both P waves (squares) and S waves (triangles). Solid symbols denote measurements made during pressurization, and open symbols denote measurements made during depressurization.

Only P-wave velocity measurements were made on water-saturated samples, since S-wave arrivals were so strongly attenuated that arrival times could not be measured accurately. Changes in P-wave velocities and the corresponding changes in porosity, determined from changes in the pore volume, are shown in Fig. 8a, b, respectively. Overall, the data show the same trend as that observed for the dry measurements, with increases in velocity of about 30 to 40% over the pressure interval 5 to 80 MPa. The saturated samples have velocities approximately 10 to 15% higher than dry samples at the same pressure, and the change in velocity is even more marked at low pressures. Again, almost identical patterns are seen in the saturated samples for heat-treated samples and in the velocity hysteresis. Crucially, the velocity increase corresponds to a porosity decrease of ~50% at 40 MPa and ~70% at 80 MPa, showing that the main deformation mechanism for water-saturated samples is inelastic pore collapse. Since the change in velocity occurs over the same pressure range as in the dry samples and the pore pressure applied was very low (1.5 MPa), it is suggested that a similar porosity decrease occurred during pressurization of the dry samples.

Taken together, we suggest that the high initial (45%) and the low final (15%) porosity of the material, the velocity hysteresis and the post-test images (Fig. 9)

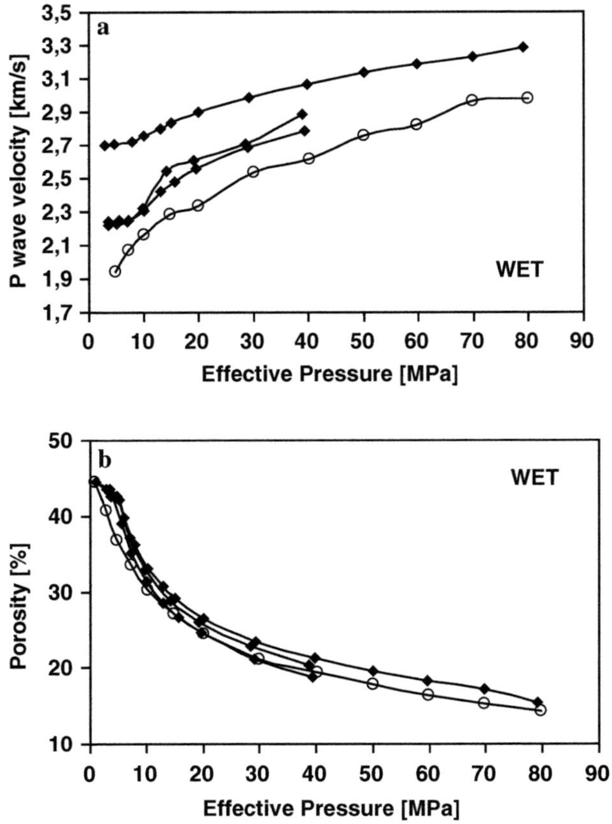


Figure 8
Plot of (a) P-wave velocities and (b) porosity as a function of effective pressure for water-saturated samples. Solid symbols denote non-heat-treated samples, and open symbols denote the sample heat-treated to 600°C.

provide strong evidence that the dominant deformation mechanism is inelastic pore collapse. This further suggests that hydrostatic pressure alone is able to drive mechanical damage of the tuff under *in situ* conditions at depth at Campi Flegrei.

### *3. Field Velocity Determinations by Seismic Tomography*

Local earthquakes that occurred at Campi Flegrei in the period 1970–1984, i.e., during the most recent episodes of unrest, provide a good database for tomographic reconstruction of seismic velocities in the depth range 0 to 4 km (Troise *et al.*, 2003; De Natale *et al.*, 1995). Seismograms of these earthquakes were recorded by both permanent (analogue) and temporary (digital) stations. A map showing the locations of the seismic stations operating during the above period, and vertical sections of

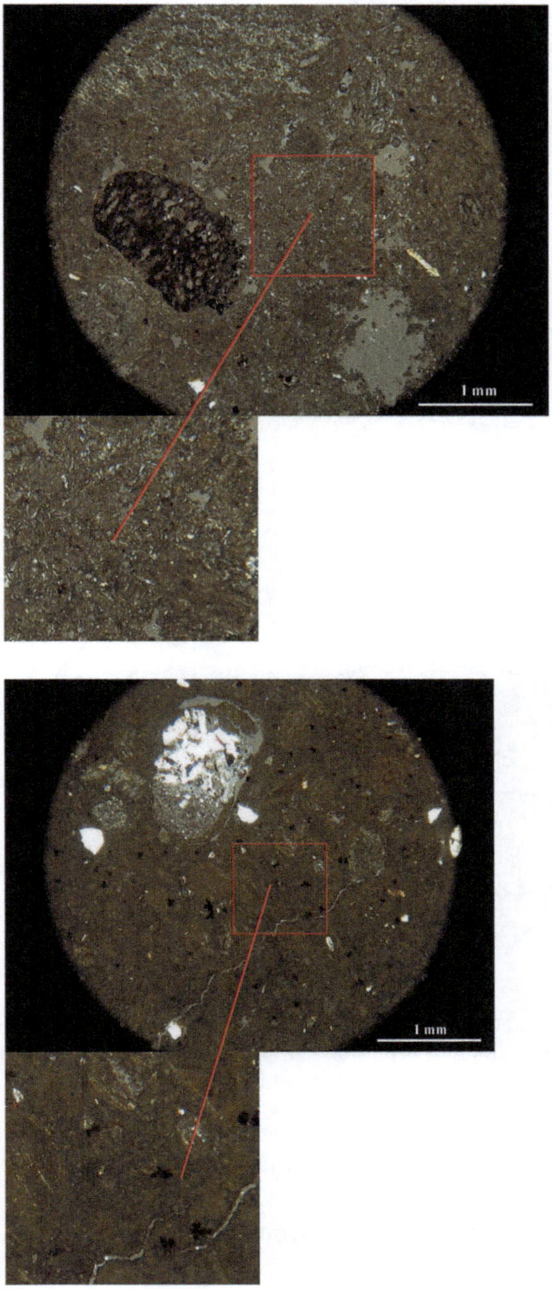


Figure 9
Photographs of a cross section of a tuff sample before (top) and after (bottom) a loading cycle to 80 MPa. Microstructure magnifications (red insets) show that inelastic pore collapse is the main deformation mechanism.

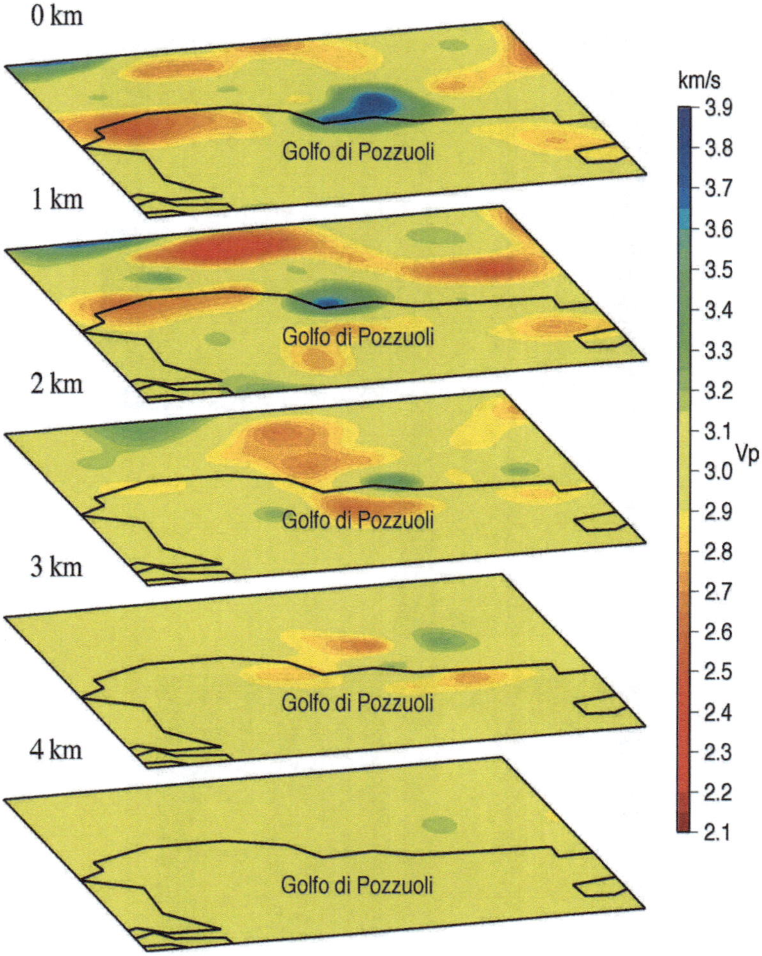


Figure 10
Tomographic map of Vp at Campi Flegrei. The (0, 0) reference point is located at the town of Pozzuoli. (a) Horizontal velocity distributions at depths of 1, 2, 3, and 4 km, and (b) vertical velocity sections with depth, oriented E-W and N-S.

earthquake locations calculated from an earlier 3-D model (PRESTI *et al.*, 2005), is presented as Figure 1.

In this study, 3757 P and S waves from a set of 451 earthquakes were inverted to obtain P- wave velocities and Vp/Vs ratio. From this full dataset, a subset of 331 earthquakes with a minimum of 8 readings and a maximum angular gap between seismic stations of 180° was selected. Three-dimensional tomographic images were generated using the inversion package SIMULPS13Q (THURBER, 1993; EBERHART-PHILIPS, 1993) which incorporates rigorous finite difference 3-D ray tracing. The starting model was taken as the best fit homogeneous velocity model. This is characterized by a Vp value 3.0 km/s, and a Vp/Vs ratio of 1.5 which gives an average RMS misfit of 2 seconds. The 3-D heterogeneous model was specified on a grid having 1 km spacing in latitude, longitude and depth. The inversion was stopped after four iterations.

We present the results for P-waves, at four depths intervals where the resolution was good enough. Results from the final Vp model are shown in Figure 10, which shows the horizontal velocity distribution at each of the four depths, and vertical cross sections of Vp, oriented N-S and E-W. The reference point of zero latitude and zero longitude is the town of Pozzuoli.

The P-wave velocity distributions characterize the structure of the inner caldera (ROSI and SBRANA, 1987) as a collapsed area filled with pyroclastic products from the recent eruptions (low velocities), evident to a depth of about 4 km and a radius of about 3 km around Pozzuoli. Since the majority of the recording stations were located within the caldera itself (Fig. 1), the resolution of the caldera rims is not as good as the inner part. However, Figure 10 does suggest a basin-like structure at depth. The average P-wave velocity in the collapsed caldera basin is around 2.9 km/s, with the highest velocities (around 3.6 km/s) occurring at the caldera center.

## 4. *Discussion and Conclusions*

When comparing seismic velocities (10 to 100 Hz) at the field scale with ultrasonic velocities (100 kHz to 1 MHz) at the laboratory scale, one has to consider both the frequency dependence and the difference in rock volume investigated.

At low frequency fluid pressure is equilibrated in pores and cracks, while at higher frequencies, the fluid has no time to flow and the fluid pressure is not equilibrated in the pores and cracks. This frequency dependent behavior produces a dispersion which is experimentally observed in isotropic media and generally called "squirt flow" mechanism (MAVKO and NUR, 1975; O'CONNELL and BUDIANSKY, 1977; THOMSEN, 1985). The degree of dispersion under saturated conditions for highly attenuating rocks, such as the Campi Flegrei tuff, has been reported to be in the range of 25 to 50% (ASTER and MEYER, 1988). Dispersion is believed to drop off rapidly to less then 10% when effective pressure is increased (ZAMORA *et al.*, 1994 and references therein), but this is strongly influenced by the density of cracks (SCHUBNEL and GUEGUEN, 2003 and references therein) in the rock. In order to quantify the dispersion process we corrected the low frequency wet velocities by applying the Biot-Gassman equations:

$$\frac{1}{K_d} - \frac{1}{K_u} = \frac{\left(\frac{1}{K_d} - \frac{1}{K_s}\right)^2}{\left(\frac{1}{K_d} - \frac{1}{K_s}\right) + \Phi\left(\frac{1}{K_f} - \frac{1}{K_s}\right)}, \tag{1}$$

where $K_d$ are the drained bulk moduli ($\sim$ equivalent to dry theoretically), $K_s$ the solid matrix bulk moduli, $K_f$ the fluid bulk moduli and $K_u$ the undrained moduli, i.e., the saturated low frequency moduli.

For the inversion, the elastic parameters were taken as $K_s = 5 \times 10^9$ Pa and $K_f = 2 \times 10^9$ Pa. $K_d$ and $K_u$ were between 3 and $9 \times 10^9$ Pa, according to the corresponding Vp.

Figure 11 shows P-wave velocities measured in wet conditions, corrected to the low frequency case via equation (1). Corrected velocities are, in general, similar to the uncorrected values, with increases ranging from 1% to 10%, in close agreement with values from ZAMORA *et al.* (1994).

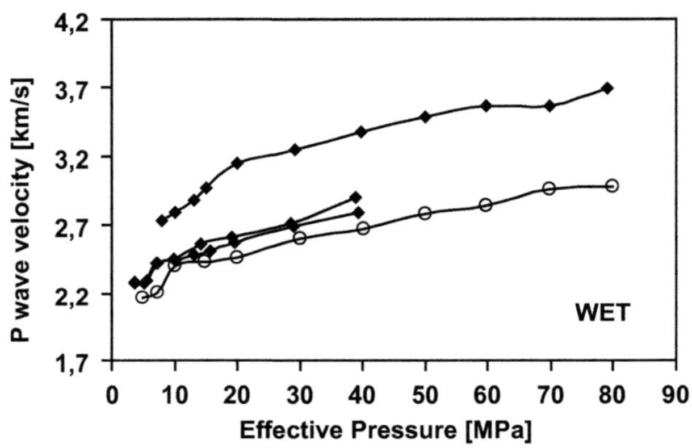


Figure 11

Plot of P-wave velocities measured in wet conditions, corrected to the low frequency case via Biot-Gassman equation (1). Solid symbols denote non-heat-treated samples, and open symbols denote the sample heat-treated to 600°C.

From these considerations, the average P-wave velocity determined for the Campi Flegrei caldera (about 2.9 km/s) is more consistent with the laboratory measurements under saturated conditions, since they are significantly higher than for dry conditions. In detail, by considering the laboratory Vp measurements performed in dry conditions (Fig. 4), one can observe that, at simulated depths of 2 km (about 40 MPa), values range from 2.3 km/s to 2.5 km/s. On the other hand, by considering Vp measurements performed in wet conditions (Fig. 8), laboratory determinations at equivalent depths range from 2.6 to 3.2 km/s, reproducing the average value of 2.9 km/s obtained by the field tomography.

Even though laboratory S-wave velocity measurements under saturated conditions are not available, the large amplitude attenuation that prohibited such measurements strongly suggests that saturated S-wave velocities are likely to be lower than those for dry samples. Hence, the velocities in water-saturated samples appear to be more representative of *in situ* rock conditions than those in dry samples, in agreement with the saturation in fluids or gas and high degree of fracturing reported for the caldera (ASTER and MEYER, 1988; O'CONNELL and BUDIANSKY, 1976; VANORIO *et al.*, 2005).

The effect of increasing hydrostatic pressure on wave velocities is strongly influenced by the geometry of the void space, and particularly by the presence of cracks and pore fluids. P- and S-wave velocities measured in the laboratory show the expected increasing trend with increase in pressure. Velocities increase by between 25 and 50% over the pressure range 5 MPa to 75 MPa. These observations, taken together with measurements of porosity decrease and images of the samples from before and after pressurization, show that pressure changes induce major damage in the tuff through inelastic pore collapse and enhanced cracking. Despite the fact that variations in velocity with increasing depth in the field-scale tomographic study are affected by the resolution of the velocity model (due to the degree of spatial clustering), an overall average increase of about 0.3 km/s is observed to occur between 2 and 4 km depth (Fig. 10). Changes in velocities at the bottom of the caldera, at about 4 km depth, are in agreement with the range expected (10–30%) from theoretical calculations (ZAMORA *et al.*, 1994).

The effect on velocities of raising the temperature appears to only become significant when it is high enough to cause some physical change. For the case of the Campi Flegrei tuff, this appears to be the volume change and generation of free water associated with the dehydration of zeolite phases, which can lead to thermal crack damage. In laboratory samples, thermal treatment to 600°C resulted in velocity decreases of about 0.3 to 0.5 km/s (i.e., about 10%) with respect to non-heat-treated samples. This is consistent with velocity changes due to increased temperature of around 10% reported by (ZAMORA *et al.*, 1994).

Overall, we have shown that structural interpretations from tomographic studies can be improved significantly if the effects of changes in rock velocities due to changes in microstructure caused by changes in environmental variables are taken into account. In particular, changes in effective pressure conditions, changes in temperature and changes in the degree of water saturation with depth all impact significantly on the propagation of elastic waves.


## *Acknowledgements*

This work has been supported by EU contract EVG1-CT-2001-00047, D'Appolonia Spa and British Council exchange programme. G. De Luca and the Servizio Sismico Nazionale, Rome kindly implemented the seismic data set. Giuseppe Mastrolorenzo, Lucia Pappalardo and Monica Piochi are thanked for fruitful discussions. Alexandre Schubnel and Derek Elsworth are acknowledged for their review, which highly improved the quality of the paper.

## References

Aster, R. C. and Meyer, R. P. (1988), *Three-dimensional velocity structure and hypocenter distribution in the Campi Flegrei caldera, Italy*, Tectonophysics *149*, 195–218.

Benson, P.M., Meredith, P.G., Platzman, E.S., and White, R.E. (2005), *Pore fabric shape anisotropy in porous sandstone and its relation to elastic and permeability anisotropy under hydrostatic pressure*, Int. J. Rock Mech. and Min. Sci. *42*, 890–899.

Bianchi, R., Cordini, A., Federico, C., Giberti, G., Lanciano, P., Pozzi, J.P., Sartoris, G., and Scandone, R. (1987), *Modeling of surface deformation in volcanic areas: The 1970–1972 and 1982–1984 crises of Campi Flegrei, Italy*, J. Geophys. Res. *92*, 14139–14150.

Bonafede, M. and Mazzanti, M. (1998), *Modelling gravity variations consistent with ground deformation in the Campi Flegrei caldera (Italy)*, J. Volcanol. Geotherm. Res. *81*, 137–157.

Bonafede, M., Dragoni, M., and Quareni, F. (1986), *Displacements and stress fields produced by a centre of dilatation and by a pressure source in a viscoelastic half-space: Application to the study of ground deformation and seismic activity at Campi Flegrei, Italy*, Geophys. J. R. Astr. Soc. *87*, 455–485.

De Gennaro, M., Incoronato, A., Mastrolorenzo, G., Adabbo, M., and Spina, G. (1999), *Depositional mechanisms and alteration processes in different types of pyroclastic deposits from Campi Flegrei volcanic field (Southern Italy)*, J. Volc. Geoth. Res. *91*, 303–320.

De Natale, G., Ferraro, A., and Virieux, J.A. (1991), *A probability method for local earthquake focal mechanisms*, Geophys. Res. Lett. *18* (4), 613–616.

De Natale, G., and Pingue, F. (1993), *Ground deformations in collapsed caldera structures*, J. Volcanol. Geotherm. Res. *57*, 19–38.

De Natale, G., Petrazzuoli, S.M., and Pingue, F. (1997), *The effect of collapse structures on ground deformations in calderas*, Geophys. Res. Lett. *24* (13), 1555–1558.

De Natale, G., Troise, C., and Pingue, F. (2001), *A mechanical fluid-dynamical model for ground movements at Campi Flegrei caldera*, J. Geodynamics *32*, 487–517.

De Natale, G., Zollo, A., Ferraro, A., and Virieux, J. (1995), *Accurate fault mechanism determinations for a 1984 earthquake swarm at Campi Flegrei caldera (Italy) during an unrest episode: Implications for volcanological research*, J. Geophys. Res. *100*, 24167–24185.

Dvorak, J. J. and Berrino, D. (1991), *Recent ground movement and seismic activity in Campi Flegrei, southern Italy: Episodic growth of a resurgent dome*, J. Geophys. Res. *96*, 2309–2323.

Eberhart-Phillips, D., *Local earthquake tomography: earthquake source regions*. In *Seismic Tomography: Theory and Practice* (eds. by Iyer, H. M. and Hirahara, K) (Chapman & Hall, London, 1993).

Gaeta, F.S., De Natale, G., Peluso, F., Mastrolorenzo, G., Castagnolo, D., Troise, C., Pingue, F., Mita, D.G., and Rossano, S. (1998), *Genesis and evolution of unrest episodes at Campi Flegrei caldera: The role of thermal fluid-dynamical processes in the geothermal system*, J. Geophys. Res. *103*, 921–933.

Gaeta, F.S., Peluso, F., Arienzo, I., Castagnolo, D., De Natale, G., Milano, G., Albanese, C., and Mita, D.G. (2003), *A physical appraisal of a new aspect of bradyseism: The miniuplifts*, J. Geophys. Res. *108*, B8, 10.1029/2002JB001913.

Mavko, G. and Nur, A. (1975), *Melt squirt in the asthenosphere*, J. Geophys. Res. *80*, 1444–1448.

O'Connell, R. and Budiansky, B. (1976), *Seismic velocities in dry and saturated cracked solids*, J. Geophys. Res. *79*, 5413–5426.

Presti, D., Troise, C., and De Natale, G. (2004), *Probabilistic location of seismic sequences in heterogeneous media*, Bull. Seismo. Soc. Am. *94* (6), 2239–2253.

Pujol, J. (2000), *Comment on Laboratory measurements of ultrasonic wave velocities in rocks from the Campi Flegrei volcanic system and their relation to other field data by M. Zamora, G. Sartoris, and W. Chelini*, J. Geophys. Res. *105* (B9), 21537–21542.

Rosi, M. and Sbrana, A. (1987), *Phlegraean Fields: Rome. Quaderni de "La Ricerca Scientifica" 9* (114), 175 pp.

Schubnel, A. and Guéguen, Y. (2003), *Dispersion and Anisotropy in Cracked Rocks*, J. Geophys. Res. *108*, 2101, doi:10.1029/2002JB001824.

Thomsen, L. (1985), *Biot-consistent elastic moduli of porous rocks: the low frequency limit*, Geophysics *50*, 2797–2807.

THURBER, C. H., *Local earthquake tomography: velocities and Vp/Vs – theory. In Seismic Tomography: Theory and Practice* (eds. Iyer H. M.and Hirahara K) (Chapman&Hall, London, 1993).

TROISE, C., DE NATALE, G., PINGUE, F., and ZOLLO, A. (1997), *A model for earthquake generation during unrest episodes at Campi Flegrei and Rabaul calderas*, Geophys. Res. Lett. *24* (13), 1575–1578.

TROISE, C., CASTAGNOLO, F., PELUSO, F., GAETA, F.S., MASTROLORENZO, G., and DE NATALE, G. (2001), *A 2D mechanical-thermal fluid-dynamical model for geothermal system at calderas: An application to Campi Flegrei*, J. Volcanol. Geotherm. Res. *109*, 1–12.

TROISE, C., PINGUE, F., and DE NATALE, G. (2003), *Coulomb stress changes at calderas: Modeling the seismicity of Campi Flegrei* (*Southern Italy*), J. Geophys. Res. *108*, B6, 2292, doi:10.1029/2002JB002006.

VANORIO, T., PRASAD, M., PATELLA, D., and NUR, A. (2002), *Ultrasonic velocity measurements in volcanic rocks: Correlation with microtexture. Geophys*, J. Internat. *149*, 22–36.

VANORIO, T., VIRIEUX, J., CAPUANO, P., and RUSSO, G. (2005), *Three-dimensional seismic tomography from P-wave and S-wave microearthquake travel times and rock physics characterization of the Campi Flegrei caldera*, J. Geophys. Res. *110*, B03201, doi:10.1029/2004JB003102.

VINCIGUERRA, S., TROVATO, C., MEREDITH, P.G., and BENSON, P.M. (2005), *Relating seismic velocities, thermal cracking and permeability in Mt. Etna and Iceland basalts*, Int. J. Rock Mech. and Min. Sci. *42*, 900–910.

ZAMORA, M., SARTORIS, G., and CHELINI, W. (1994), *Laboratory measurements of ultrasonic wave velocities in rocks from the Campi Flegrei volcanic system and their relation to other field data*, J. Geophys. Res. *99*, B7, 13553–13562.

ZAMORA, M., SARTORIS, G., and CHELINI, W. (2001), *Reply to comment by Jose Pujol on "Laboratory measurements of ultrasonic wave velocities in rocks from the Campi Flegrei volcanic system and their relation to other field data"*, J. Geophys. Res. *106*, B2, 2163–2166.

(Received October 21, 2005, revised November 21, 2005, accepted November 22, 2005)

Pure appl. geophys. 163 (2006) 2223–2242
0033–4553/06/102223–20
DOI 10.1007/s00024-006-0119-x



Pure and Applied Geophysics

# Spatial Scaling of Effective Modulus and Correlation of Deformation Near the Critical Point of Fracturing

Kes Heffer[1] and Peter King[2]

*Abstract*—Many observations point to the lithosphere being metastable and close to a critical mechanical point. Exercises in modelling deformation, past or present, across subsurface reservoirs need to take account of this criticality in an efficient way. Using a renormalization technique, the spatial scaling of effective elastic modulus is derived for 2-D and 3-D bodies close to the critical point of through-going fracturing. The resulting exponent, $d_\mu$, of spatial scaling of effective modulus with size, $L^{d_\mu}$, takes the values $\sim -2.5$ and $-4.2$ in two- and three-dimensional space, respectively. The exponents are compatible with those for scaling of effective modulus with fracture density near the percolation threshold determined by other workers from numerical experiments; the high absolute values are also approximately consistent with empirical data from a) fluctuations in depth of a seismic surface; b) '$1/k$' scaling of heterogeneities observed in one-dimensional well-log samples; c) spatial correlation of slip displacements induced by water injection. The effective modulus scaling modifies the spatial correlation of components of displacement or strain for a domain close to the critical point of fracturing. This correlation function has been used to geostatistically interpolate components of the strain tensor across subsurface reservoirs with the prime purpose of predicting fracture densities between drilled wells. Simulations of strain distributions appear realistic and can be conditioned to surface depths and observations at wells of fracture-related information such as densities and orientations, welltest permeabilities, changes in well-test permeabilities, etc.

**Key words:** Effective modulus, critical point, correlation function, renormalization, fractures, scaling.

## *Introduction*

### *Fracture Modelling*

One of the prime objectives of current research in geological modelling is to develop realistic methods for prediction of fracture configurations and conductivities, particularly by interpolation between data available at sparse well locations. However, the detail with which these need to be interpreted in order to confidently predict well productivities or flow properties for, say, oilfield development planning is fortunately not always necessarily at the individual fracture scale. In many cases it

[1] Institute of Petroleum Engineering, Heriot Watt University, Riccarton, Edinburgh, EH14 4AS, UK. E-mail: kes.heffer@pet.hw.ac.uk; kayness@nildram.co.uk
[2] Department of Earth Science and Engineering, Imperial College, Exhibition Road, London, SW7 2AZ, UK.

is expected that local fracture properties might be reasonably predicted from local values of the components of the total strain tensor.

For calibration, estimates of local values of the strain tensor can be made from the fractures observed in cores or image logs from wells. However, to do so requires measurement of fracture displacements: apertures or shear displacements (or in the case of stylolites, horizontal or vertical, the degree of dissolution). These are difficult to determine and so, even with plentiful data on fractures at wells, the local estimate of strain is approximate only. Existing estimates of fracture permeability under modern-day stress conditions are also often available from well tests, if the contribution to permeability due to the matrix rock can be backed out from the test permeabilities. Although only scalar permeabilities are thereby available, they can be used as additional conditioning data for modelling (and might well be the best available data in some cases).

Curvature analysis of surfaces is a technique which has been used for several decades in the hydrocarbon industry as a guide to interpolating fracture densities (and strike orientations if the principal axes of the full curvature tensor are analyzed). However, the success rate of this technique has not generally been high. One reason for this is that curvature analysis only examines the dilatational component of strain of the rock. The shear components of deformation are ignored. For dip-slip shear displacements on (micro-)faults, a high local curvature of an intersected surface would be interpreted but which is not necessarily associated with high densities of extension fractures; for strike-slip displacements, no local curvature of the surface would be interpreted, and yet there may well be extensional strain and conductive fractures due to lateral displacements. In order to interpret the full 3-D strain tensor, it is necessary to make an estimate of the lateral (horizontal) displacements. These are invisible to conventional seismic surveys (with the exception of modern anisotropy surveys).

Interpolation of strain data between wells can be better effected with the 2-point correlation function. Although using just 1- and 2-point statistics neglects the likely multifractal nature of structural features (e.g., COWIE *et al.*, 1995; OUILLON *et al.*, 1996; MARSAN and BEAN, 2003), conditioning to measured well data should ameliorate this potential distortion. It is the purpose of this paper to describe one form of spatial correlation that draws on basic elastic theory, but modified to account for the presence of fractures. The correlation involves the spatial scaling of effective modulus near a critical point, which is newly developed in this paper using real-space renormalization.

HEFFER *et al.* (1999) proposed such a geostatistical method for spatially interpolating the strain field between conditioning data available from local strain estimates based upon observations of fractures at well locations. The Appendix describes an adaptation of LANDAU and LIFSHITZ (1975) to derive the spectral density function of the 2-point correlation function for elastic displacements, in which non-interacting cracks are assumed to give rise to a Boltzmann or power-law distribution of elastic strain energy: it is given in terms of the Fourier components of the displacement vector $\mathbf{u(k)}$ as:

$$\langle u_i(\mathbf{k}).u_j(\mathbf{k})\rangle = \frac{\langle E\rangle}{2\mu}k^{-2}\left[\delta_{ij} - \frac{1}{2(1-\upsilon)}\frac{k_i k_j}{k^2}\right], \tag{1}$$

where k is the wavenumber, or spatial frequency, $\mu$ is the shear modulus of elasticity, $v$ is Poisson's ratio and $\langle E \rangle$ is the average strain energy. The eigenvectors of the correlation matrix at each wavenumber can be shown to correspond to the three modes of deformation: 1 dilatational and 2 shear. [Note that, although the elasticity or stiffness tensor is in general associated with six eigenstrains (e.g., KELVIN, 1856 or MEHRABADI and COWIN, 1990), any three of those eigenstrains are redundant in a deformed body, because of the constraints of the Saint-Venant compatability equations (e.g., JAEGER and COOK, 1979, p. 44), which ensure that the line integral of strain along any arbitrary line between two points gives the same change in displacement values.] Expressions for the correlation of strain components can be derived from the correlations of the displacements by differentiation.

During the derivation of spatial covariances of displacements in the Appendix fractures were taken into account in the limited sense of assuming the frequency distribution of strain energy that would accompany a general configuration of fractures, but that energy was calculated on an elastic basis, without correcting the elastic moduli for the presence of fractures. Fractures reduce the strain energy in the rock. Also, as crack densities increase, the strain energies associated with crack-crack interactions become important. Nevertheless, equation (1) was an improvement upon curvature analysis (to which it reduces in the case of no shear deformation, and only conditioning to a single surface).

### *Lithosphere at a Point of Self-organized Criticality*

Since the early 1980s there has been growing evidence for the concept of self-organized criticality (or near-criticality) as a general model of deformation of the lithosphere, in which the continuous dissipative transfer of energy (strain, hydraulic, thermal, chemical, etc.) results in percolating paths of metastable structures (e.g., MAIN, 1996; BAK, 1997). Such a system is characterized by power-law distributions, metastability (or large susceptibility to perturbations), and long-range correlations. In order therefore to develop a full theory of strain correlation, we need to take into account the following factors:

- near-criticality, at which fracture interactions are vital,
- large range of fracture sizes,
- anisotropy of fractures,
- non-random configurations of fractures.

One way of dealing with the contribution to energy from the fractures is to retain the same form as equation (1) for the spatial covariance of displacements, but to modify the scalar elastic modulus to an effective value that takes into account the presence of fractures.

Renormalization techniques were developed specifically for analyzing critical phenomena in systems over a large range of scales. They overcome the problem that is associated with mean-field approaches by incorporating configurations that depart from the average, but which significantly affect the scaling. Renormalization in real space was first introduced by KADANOFF (1966) and is given extensive discussion in BINNEY *et al.* (1992). Following this method, a means of tackling the large range in fracture sizes is to treat them in a hierarchy of scales, in which the effective modulus from incorporating fractures at one scale is used as the 'matrix' modulus for the next scale up. This will lead to a scale-dependence of the effective modulus, which at the critical point for through-going deformation, where there is no characteristic length scale, will approach a power-law. Renormalization assumes that the energy of a system near the critical point has the same functional form at all scales. In our case, strain energy = effective modulus × (strain)$^2$; or, in general tensorial form, $E = \frac{1}{2}e_{ij}\mu^e_{ijkl}e_{kl}$.

At each stage of the renormalization, use is still made of a large background of mean-field approaches to effective moduli produced, mainly by workers studying acoustic wave properties in fractured rock. A summary of those approaches is given in the next section.

*Existing Literature on the Effect of Fractures on Elastic Moduli*

KEMENY and COOK (1986) and SAYERS and KACHANOV (1991) give good summaries of developments in this topic, and the following paraphrases those. For higher crack densities, when crack interactions cannot be neglected, O'CONNELL and BUDIANSKY (1974) and BUDIANSKY and O'CONNELL (1976) proposed a self-consistent scheme for the calculation of the elastic moduli for random crack orientation statistics. In this method, the effect of crack interactions is included by assuming that each crack is embedded in a medium with the effective stiffness of the cracked body. As pointed out by BRUNER (1976) and HENYEY and POMPHREY (1982), this self-consistent scheme may overestimate the crack interactions. They proposed a similar scheme, that only differed by increasing the crack density in small steps after each of which the elastic properties are recalculated. This is referred to as the 'differential' scheme, for which ZIMMERMAN (1985) gave the exact solution, and ZIMMERMAN (1991a) provided detailed discussion. KEMENY and COOK (1986) incorporated in this approach strong interactions between fractures through the device of 'external' cracks (two parallel notches in opposite sides of a domain, whose tips approach one other). This modification therefore caters for large fracture densities, although still assuming random orientations and consequent isotropic effective moduli. HUDSON (1980, 1981, 1986) has given results for both randomly orientated and parallel cracks that are correct to second order in the crack density. His results however, are restricted to small crack densities and give an unrealistic increase of the moduli with crack density for moderate crack densities, as shown by SAYERS and KACHANOV (1991). SAYERS and KACHANOV (1991) themselves presented a simple scheme for finding the effective elastic properties of solids for arbitrary orientation statistics at finite crack densities, based on a tensorial transformation of the effective elastic constants. This transformation, known for randomly oriented penny-shaped cracks was extended, to the case of arbitrary orientation statistics, through the use of a second-order crack density tensor characterizing the averaged geometry of the crack array. SHEN and LI (2004) included in their review of methods for calculating effective moduli further variations that have been devised by other authors to cope with crack interactions.

### *Real-space Renormalization Approach*

The renormalization method iteratively calculates average values of the parameters of the system (the effective modulus in our case) in terms of their values at the previous scale, taking into account the discrete probabilities of all (or most of) the possible configurations of the system (BINNEY *et al.*, 1992).

A possible route is to perform a finite-element geomechanical simulation of a very finely gridded domain loaded close to failure, and to coarsen the variables of the grid in a Monte Carlo renormalization scheme (BINNEY *et al.*, 1992, section 5.10), in which average values of various groupings of a state variable (stress or strain components) comprising the Hamiltonian (energy) on the original grid are related to their corresponding values on the coarsened grid through a matrix. The exponents of the scaling characteristics of the model parameters (i.e., the components of effective moduli) would then be related to the eigenvalues of that matrix.

However, as with all Monte Carlo renormalization approaches, a major problem is to ensure an adequate sampling of all possible configurations: This would require sophisticated treatment of the boundary conditions for the finite element model. The renormalization approach described below assumes a much simpler model, which allows the full space of configurations to be examined.

ALLÈGRE *et al.* (1982) made use of a renormalization group theory in real space in developing a model of fracturing. In this, fractures at one scale were combined under a simple rule for deciding whether fracturing existed at a renormalized scale. The non-linear equation linking the probability of fracturing at the smaller scale with the probability at the larger scale shows a critical point at which the probability of fracturing is scale-independent. When probability of small fractures occurring is below that critical point, then fracturing is limited in scale; with probability at or above the critical point, then fracturing extends to all scales. SMALLEY *et al.* (1985), TURCOTTE (1986) and ALLÈGRE and LE MOUËL (1994) also used the renormalization group approach to model the fracture behavior of rocks. ALLÈGRE and LE MOUËL (1994) extended the work of ALLÈGRE *et al.* (1982) to incorporate interactions between neighboring fractures, which can also be at different angles to each other. In this present approach, the method of ALLÈGRE *et al.* (1982) is supplemented by calculating not only the probability of fracture from one scale to the next, but also the average effective modulus of each configuration. Note that in this paper the elastic modulus, $\mu$, is assumed to be a scalar for the purposes of calculating its spatial scaling, and can be taken as either the Young's or shear modulus. This approach could be extended to the renormalization of anisotropic elastic tensors via the multi-dimensional model of fracturing assumed in ALLÈGRE and LE MOUËL (1994). It is also possible that the renormalization equations for the effective moduli in the larger cell could be formulated directly from their values in each of the sub-cells, without invoking the effective medium approach; KING (1989) developed such renormalization equations for averaging heterogeneous permeabilities.

Figure 1 depicts the concept of renormalizing the probability of fracture and the effective modulus, respectively. Renormalization provides a relationship between parameters of a small-scale cell and those of a larger-scale cell, the latter being defined as a block of contiguous smaller-scale cells (e.g., $2 \times 2 \times 2$ block). The renormalization procedure, as set out by ALLÈGRE *et al.* (1982) for fracturing is as follows.

With a given probability of fracture at the small-scale cell size, each possible configuration of fractured and unfractured cells in a larger-scale cell is considered. In this study the small-scale cells were assumed to be independent with regard to the probability of fracture, although interactions between neighboring fractures could be introduced by modification of the probabilities in a manner along the lines of that used by ALLÈGRE and LE MOUËL (1994). It should also be noted that the effective medium methods of calculating effective moduli, to some extent, take into account an energy of interaction of fractures (for more comment on this point see the 'Results' section below).

For each configuration of small-scale cells, the probability of the occurrence of fracture of the larger-scale cell was calculated, using some rule (e.g., in a $3 \times 3$ two-dimensional grid, alignment of three small-scale fractures). Renormalization takes the process to the next stage of iteration at the next scale of cells. The overall

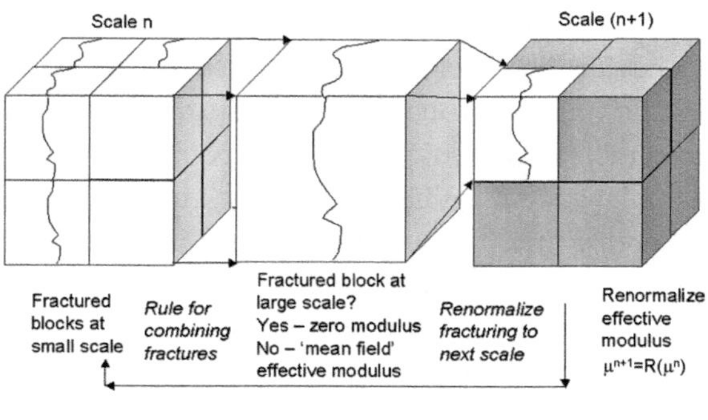


Figure 1
Renormalization technique for effective modulus. Adaptation of ALLÈGRE, LE MOUËL and PROVOST (1982). For explanation see text.

relationship between probability of fracture of the larger-scale cell and that of the small-scale cells is derived by taking into account all possible configurations of fracturing. The relationship generally has a (non-trivial) critical point at which the probability of fracturing is constant across all scales. During a sequence of increasing strain in a domain of rock, fracturing is limited in scale until the density reaches the critical point, at which there is catastrophic failure.

Renormalization of the average effective modulus of the fractured system is formulated in parallel with the renormalization of fracturing. At each scale the mean effective modulus is calculated, considering all possible configurations of small-scale fractures, governed overall by the critical density. No correlation in the fractures is assumed, and all knowledge of where fractures existed at the previous scale is discarded at each renormalization. This is in line with the approach used by ALLÈGRE *et al.* (1982). The background 'matrix' modulus is taken to be the mean effective modulus calculated at the previous scale. The details are as follows:

a) If the larger-scale cell of the configuration is deemed under the assumed rule to be 'fractured', its effective modulus at that scale is taken as zero. This ignores any stiffness that might be provided to shear fractures by friction, although, again, this might be added in future studies.

b) If the larger-scale cell of the configuration is 'unfractured', its effective modulus, $\mu^{n+1}$, is calculated from that at the small-scale average effective modulus, $\mu^{n}$, by recourse to one of the effective medium techniques that have been described earlier. Note that the mean-field approach is being used only for the approximation to the interaction of the lower density configurations of fractures in 'unfractured' large-scale cells; fracture coalescences in higher density cells are catered for by the more drastic reductions in effective modulus executed in a). Approximations such as this are common in renormalization schemes, which are affected more by configurational geometries than by details of interactions. The relationship for effective medium theories is $\mu^{n+1} = a.\mu^{n}$, where a takes the following forms in terms of the fracture density, $\rho$, appropriate for the specific configuration being considered:

$a = (1+b^*\rho)^{-1}$ for non-interacting fractures

$a = (1-b^*\rho)$ for the self-consistent approach of O'CONNELL and BUDIANSKY (1974) and BUDIANSKY and O'CONNELL (1976)

$a = \exp(-b^*\rho)$ for the differential scheme of BRUNER (1976), HENYEY and POMPHREY (1982), and ZIMMERMAN (1985).

The geometrical constant, b, as outlined in KEMENY and COOK (1986), is, for a unidirectional set of fractures:

$$b = 2\pi(1-\upsilon^2)\left(\frac{c}{L}\right)^2 \text{ in 2-dimensions, and } \ldots$$

$$b = \frac{16}{3}(1-\upsilon^2)\left(\frac{c}{L}\right)^3 \text{ in 3-dimensions,}$$

where $c/L$ is the radius of each fracture as a proportion of the cell dimension at any scale, and $\nu$ is Poisson's ratio (assumed constant across scales in this work). The effective modulus associated with each possible configuration, i, of fracturing is weighted by the probability, $p_i$, of occurrence of that configuration at the critical fracture density, and the mean effective modulus at the new scale obtained from the weighted sum over all possible configurations: $\langle \mu^{n+1} \rangle = \sum_i p_i^{n+1} \mu_i^{n+1}$. The spatial scaling of modulus is obtained from the ratio between mean effective moduli at consecutive scales. Recognizing that there is no characteristic scale at the critical point of fracturing, the mean effective modulus will be a power-law function of length scale, $L$; i.e., $\langle \mu^n \rangle \sim (\mathrm{L}^n)^{d_\mu}$. The exponent, $d_\mu$, is therefore obtained from the ratio between consecutive scales of the mean effective modulus as:

$$d_\mu = \log \left( \frac{\langle \mu^{n+1} \rangle}{\langle \mu^n \rangle} \right) / \log \left( \frac{L^{n+1}}{L^n} \right).$$

## *Results*

Table 1 gives the results of the above algorithm. The various cases analyzed have included both two- and three-dimensional configurations, and different rules for determining whether larger-scale cells are fractured. As expected, the differential scheme produces exponents which are intermediate between those of non-interacting crack and self-consistent schemes. Previous numerical results and experimental data indicate that the self-consistent approach overemphasizes the interaction between cracks, and that moduli of solids with cracks or voids lie between non-interacting and differential calculations (SHEN and YI, 2001; ZIMMERMAN, 1991b; HUANG *et al.*, 1996; RODIN and HWANG, 1991; WANG *et al.*, 2000). As pointed out by an anonymous reviewer of this paper, SHEN and LI (2004) obtained numerical results for

Table 1

*Results of real-space renormalization of effective modulus for various embedding dimensions; number of small-scale cells comprising a large-scale cell; rules for determining failure at the larger scale from configuration of fractures in the small-scale cells; and effective medium model chosen for 'unfailed' large-scale cells. The values of the modulus scaling exponent, $d_\mu$, considered most representative are high-lighted in bold italics*

| Case | Dimensions | No. of cells | Failure rule | Critical prob. | Scaling exponent for modulus, $d_\mu$ | | |
|---|---|---|---|---|---|---|---|
| | | | | | Non-interacting | Self-consistent | Differential |
| A | 2 | 2 × 2 | 2 in a plane | 0.233 | −0.616 | −0.716 | −0.654 |
| B | 2 | 2 × 2 | 2 in a line | 0.618 | ***−2.435*** | −2.894 | ***−2.572*** |
| C | 2 | 3 × 3 | 3 in a line | 0.682 | −1.573 | −2.529 | −1.764 |
| D | 3 | 2 × 2 × 2 | 3 in a plane | 0.393 | −0.797 | −0.862 | −0.824 |
| E | 3 | 2 × 2 × 2 | 4 in a plane | 0.921 | ***−4.199*** | −4.542 | ***−4.316*** |

random sizes of cracks that were very close to the predictions of the differential method, supporting the assertion by SALGANIK (1973) that the differential method should be more accurate as the crack size distribution becomes broader. However, in the renormalization method of this paper the effective modulus of the system is calculated at each stage with cracks of only one scale. It is therefore a moot point, left for future investigation, as to whether SHEN and LI'S conclusion is applicable here, and we restrict ourselves to concluding that the exponents from the non-interacting cracks and differential schemes are more representative.

The range of exponents resulting from the different configurations and schemes can be roughly interpreted, favoring the most appropriate failure rules, to indicate that the spatial scaling of average effective modulus has exponent, $d_\mu$ , of about −2.5 for two dimensions and about −4.2 for three dimensions. The value for three dimensions is in line with those obtained by previous theoretical investigations, and the value for two dimensions is within the range of previous theoretical values. The high absolute values of the exponents indicate very rapid softening of effective modulus for larger and larger domains. Remember that this only applies at, or near, the critical point of fracturing. It implies that displacements are more closely correlated (or anti-correlated, depending on orientation) at larger separation displacements. Although initially this might seem counter-intuitive, it is consistent with coherent displacements at either end of a fracture, or fault: the larger the fracture, the greater the displacements involved. Also note that, because the scaling of modulus is applicable only near to the critical point of inelastic deformation, it does not imply any dispersion in seismic body waves (as would a scale-dependency in modulus for elastic deformation).

*Consistency with Previous Investigations of Scaling of Modulus*

CHAKRABATI and BENGUIGUI (1997) discuss from a theoretical viewpoint the scaling of modulus close to failure. They look (their section 3.4) at a disordered solid modelled by a percolation model. They consider a random bond network with both central (along bond) and bond-bending (changing bond angles) force constants that govern the elastic energy due to bond stretching and bond-bending, respectively. When the bond occupation concentration $p$ is below the percolation threshold, $p_c$, the elastic modulus is of course zero. As p rises above $p_c$, the elastic modulus grows according to the power-law: $\mu \sim (p - p_c)^{T_e} for\; p > p_c$, where $T_e$ is called the elastic exponent. Extensive numerical studies have ascertained that $T_e \sim 3.96$ and 3.75 in 2 and 3 dimensions, respectively. Experimental results also compare well with the numerically estimated exponents. FENG (1985) reviews the values of $T_e$ that have been found by other workers as in the range 3.2–3.7 in two dimensions and 3.6 in three dimensions. KANTOR (1985) deduced theoretical upper and lower bounds on the elastic modulus exponent, $T_e$, for a central and bond-bending model: these are close to 4 for any dimension.

Following ARBABI and SAHIMI (1988) we can obtain a length scaling relationship for modulus by a finite-size scaling argument in which we consider the scaling behavior of the correlation length close to the percolation threshold:

$\xi(p) \sim |p - p_c|^{-\upsilon}$ for $|p - p_c| << p_c$, where $\upsilon$ has values 4/3 and 0.88 in 2 and 3 dimensions, respectively. The correlation length is the average over all clusters of the RMS distance between pairs of sites belonging to the same finite cluster. Therefore, a domain of size $L$ in a system close to the percolation threshold will have behavior indistinguishable on average from a system at the threshold, $(\xi \sim L)$, if the corresponding deviation of the bond occupation concentration from the threshold value in the parent system is $|p - p_c| \sim L^{-1/\upsilon}$, and so we obtain for the modulus scaling at the threshold: $\mu \sim L^{-T_e/\upsilon}$. The exponent for modulus spatial scaling $-T_e/\upsilon$, has the values 2.97 and 4.26 in $d = 2$ and 3 dimensions, respectively.

GUYON *et al.* (1989) review the results of applying percolation theory to a central-force lattice, which comprises a 2-D triangular lattice of springs rotating freely at their nodes. In the same random dilution problem in which a proportion (1-$p$) of randomly distributed bonds are missing, they state the same power-law for elastic modulus, but above the rigidity threshold, $p_r = 0.642$, which is higher than the normal connectivity percolation threshold ($p_c = 0.5$), because simple connectivity does not ensure the rigidity of the structure in this case. Nevertheless their numerically determined exponent ratio: $T_{ce}/\upsilon_{ce} = 3.0 \pm 0.4$, is still very close to that for bond-bending percolation. However, as noted by FENG (1985), two other estimates of $T_{ce}$ by FENG and SEN (1984) and LEMIEUX *et al.* (1985) do not agree. Additionally CHAKRABATI and BENGUIGUI (1997) quote a value for $T_{ce}/\upsilon_{ce} = 1.12 \pm 0.05$ in two dimensions. ARBABI and SAHIMI (1988) found $T_{ce}/\upsilon_{ce} = 1.14 - 1.42$ from a two-dimensional model with central forces and proposed that there may not be universality of elastic exponents for such models.

ZHOU and SHENG (1993) investigated the effect of isolated pockets of fluid in a 2-D continuum elastic solid. The changes in shape of the fluid pockets during shear means that the fluid exerts an extra stiffening effect to the shear modulus, which was found to scale with an exponent of $0.91 \pm 0.02$, rather than $2.6 \pm 0.3$ determined for a solid-vacuum continuum. The authors note that the stiffening effect of the fluid would be minimal if liquid flow is possible.

So, whilst there are some variations in determined values of the elastic modulus scaling exponent, the consensus with the bond-bending model is for high values (3 or 4): modulus decays very rapidly with scale in a system near criticality.

### *Incorporation of Scale Dependency of Effective Modulus into Correlations of Deformation*

The value of effective modulus appropriate to a spectral component of fluctuation with wavenumber, $k$, is that associated with the lengthscale $\sim 1/k$. So scaling of modulus is $\mu(k) \sim k^{T_e/\upsilon}$. Note that we do not take the Fourier transform over space of $L^{-T_e/\upsilon}$, since we are considering the general scaling of moduli for various sizes of domain, or various wavenumbers in Fourier space, rather than the specific variation of modulus with position over a given domain. We can now make an heuristic adaptation of the elastic theory of correlation function derived in the Appendix. Since the elastic theory is developed for each wavenumber, then it can be modified to the inelastic case by substitution of the appropriate effective modulus for each wavenumber, $\mu$ ($k$). The shear modulus appears in the denominator of the expression for the covariance of displacements, so that the exponent of the magnitude of wavenumber, $k$, in that expression, when amended for inelastic behavior, changes from –2 to $-(2 + T_e/\upsilon)\sim -5$ to –6 in three-dimensions. As noted by KALLESTAD (1998), the Fourier transform of $r^{-\lambda}$ in D-dimensional space is given by:

$$F\{r^{-\lambda}\}(k) = \pi^{-\lambda-D/2}\frac{\Gamma\left(\frac{\lambda+D}{2}\right)}{\Gamma\left(\frac{-\lambda}{2}\right)}k^{\lambda-D},$$

where Γ () is the gamma function.

So $k^{-6}$ ($\lambda = -3$) corresponds to $r^3$ in three-dimensional space. Also noted by KALLESTAD (1998), the function $r^\lambda$, where $\lambda > -D$, is a positive-definite function in D-dimensional space. Therefore, the exponent of –6 for the wavenumber is the minimum value for ensuring a valid correlation function in real space.

*Sample Simulations*

Figure 2 shows a progression of simulations of maximum principal horizontal strain in a case example, which is constrained to top and bottom surfaces of a

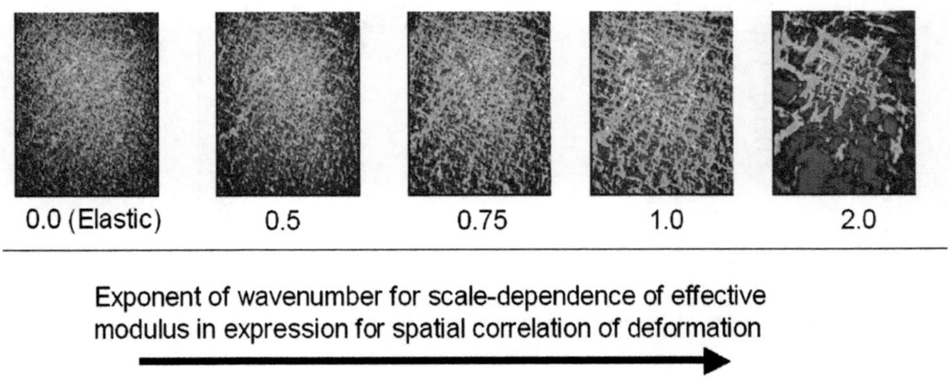


Figure 2

Simulations of the maximum principal strain field across a large reservoir, constrained to top and bottom surfaces, and assuming progressively larger exponents in the scale-dependence power-law of the elastic modulus. Spatial correlation increases from left to right with larger exponents – realism also seems to increase.

formation. From left to right the simulations have used progressively higher values of the exponent of wavenumber in the correlation of displacements. As expected, it is seen that higher values of the exponent lead to greater spatial correlation in trends of high or low strain. The longer correlation lengths look qualitatively more realistic. Variations in the thickness of the formation are converted by the application of the algorithm into laterally consistent lineations in horizontal strain.

## *Comparisons with Empirical Data*

### *Autocorrelation of Stratigraphical Surfaces*

If we assume that the current depth contours of a stratigraphical surface represent the vertical displacements of an originally flat surface, $u_z(r)$, then the spectral power $u_z(k).u_z(-k)$ should reveal an empirical estimate of the exponent of the wavenumber.

Figure 3 shows the full spectral power of a sample surface, digitized on a 400 × 375 grid, plotted on a log-log plot against the magnitude of the wavenumber, $k$. There is a wide scatter to the points, but the average trend picked by eye through the scatter has a slope of about –5. Athough the effects of depositional and erosional variations in surface levels need to be considered, these data seem to confirm the strong dependence of the effective elastic modulus upon the scale (or wavenumber).

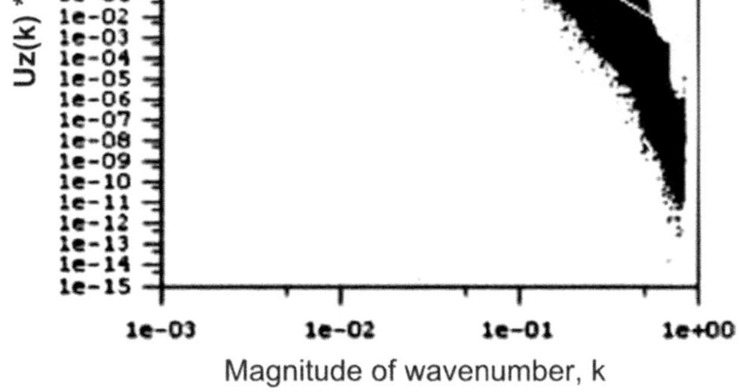


Figure 3
Spectral power of the depth to a field seismic surface on a log-log plot against wavenumber of the areal discretization ($Uz(k)$ is the Fourier transform of the vertical depth to the surface across its areal domain, for wavenumber magnitude $k$).

*1/k Scaling of Heterogeneities in 1-D Samples*

The scaling of correlation of displacements near the critical point of inelastic three-dimensional deformation, derived above as $k^{-5}$ to $k^{-6}$, implies that the correlation of three-dimensional inelastic strain, the first derivative of displacement, near the critical point will scale as $k^{-3}$ to $k^{-4}$. In real three-dimensional space, the associated correlation of strains varies with lag distance, $r$, as $r^0$ (or $\log(r)$) to $r$. A one-dimensional sample through such correlated strain with these forms will have the same scaling in real space, of which the spectral power in Fourier space will be $k^{-1}$ to $k^{-2}$. This has a possible relationship with widespread observations of so-called 'flicker', or 'pink', 'noise' (also referred to as '$1/f$ noise', or, with $k$ rather than $f$ as the scalar spatial frequency, '$1/k$ noise') in one-dimensional well logs of different variables (porosity, density, permeability, etc.) e.g., LEARY (2002). In terms of wavenumber, $k$, the spectral power densities of the heterogeneities show power-law behavior $S(k) \sim 1/k^{\beta}$ where $\beta \sim 1.0$ to 1.6. This hints strongly at the possibility that the basis of this behavior is the scaling of fluctuations in strain. The link is most plausible for measurements of fracture porosity in fractured crystalline rocks. In sedimentary rocks it is possible that palaeo-strains provided an original framework that was later exploited and amplified by diagenetic trends; alternatively, the geomorphology associated with palaeo-structural trends may have influenced the actual depositional patterns directly. Here comparison is made with the Landau-Ginzburg 'meta-model', commonly used for critical phenomena in thermodynamic equilibrium. Fluctuations in the order parameter of a three-dimensional process are associated with spectral densities that vary as $1/k^{2-\eta}$, where, in the Landau-Ginzburg model, $\eta$ takes a small value $\sim$0.1 (BINNEY *et al.*, 1992). If we take the order parameter of deformation to be the total strain, fluctuations would have spectral densities $\sim k^{-2}$ from this model, corresponding to a spectral density for displacements $\sim k^{-4}$, in contrast with $\sim k^{-2}$ for the case of non-interacting cracks in equation (1).

In contrast, the value of the correlation exponent, $\eta$, that corresponds to $1/k$ scaling of one-dimensional samples (or $k^{-3}$ scaling of three-dimensional fields), is $\eta = -1$. In real space renormalization $\eta = -1$ is also associated with the 'low temperature' value of the normalization parameter, $\omega_c = (d + 2 - \eta)/(2d) = 1$ (BINNEY *et al.*, 1992, section 5.4.1). It implies that renormalization in real space of the order parameter takes the form of a simple arithmetic averaging over the smaller cells comprising a renormalized cell. Although $\omega_c$ cannot be determined for a linear real-space renormalization scheme, arithmetic averaging is as expected if we take the components of the conventional (infinitesimal) strain tensor as the order parameters of inelastic deformation near the critical point. It is conjectured that there may be better order parameters than strain that scale more closely, with different values of scaling exponent and of $\eta$ and $\omega_c$. However, the familiarity of strain and the empirically observed scaling of the porosity of fractured crystalline rock imply that

Table 2

*Summary of various models in various dimensions for correlation of strain fluctuations, sampled in 1, 2 or 3 dimensions, with their associated scaling exponents*

| Sampling ► dimension | 1d sample | | | | | | 2d | 3d sample | | | | | | |
|---|---|---|---|---|---|---|---|---|---|---|---|---|---|---|
| Model ▼ | Real | FT | H | D | β | $\eta$ | FT | Real | FT | H | D | β | $\eta$ | $\omega_c$ |
| Elastic | $r^{-3}$ | $k^{2}$ | $-\frac{3}{2}$ | 3.5 | −2 | 2 | k | $r^{-3}$ | $k^{0}$ | $-\frac{3}{2}$ | 5.5 | −2 | 2 | $\frac{1}{2}$ |
| Critical point - 3D space process | $r^{-(1+\eta)}$ | $k^{\eta}$ | $-\frac{(1+\eta)}{2}$ | $\frac{(5+\eta)}{2}$ | $-\eta$ | $\eta$ | $k^{\eta-1}$ | $r^{-(1+\eta)}$ | $k^{\eta-2}$ | $-\frac{(1+\eta)}{2}$ | $\frac{(9+\eta)}{2}$ | $-\eta$ | $\eta$ | $\frac{(5-\eta)}{6}$ |
| Critical point - 1D space process | $r^{(1-\eta)}$ | $k^{\eta-2}$ | $\frac{(1-\eta)}{2}$ | $\frac{(3+\eta)}{2}$ | $2-\eta$ | $\eta$ | | | | | | | | |
| Critical point mean-field model* - 3D space process | $r^{-1}$ | $k^{0}$ | $-\frac{1}{2}$ | 2.5 | 0 | ˜0 | $k^{-1}$ | $r^{-1}$ | $k^{-2}$ | $-\frac{1}{2}$ | 4.5 | 0 | ˜0 | $\frac{5}{6}$ |
| fBm '1/k' | Log(r) | $k^{-1}$ | 0 | 2.0 | 1 | −1 | $k^{-2}$ | Log(r) | $k^{-3}$ | 0 | 4.0 | 1 | −1 | 1 |
| | r | $k^{-2}$ | $\frac{1}{2}$ | 1.5 | 2 | −2 | $k^{-3}$ | r | $k^{-4}$ | $\frac{1}{2}$ | 3.5 | 2 | −2 | $\frac{7}{6}$ |
| | $r^{2}$ | $k^{-3}$ | 1 | 1.0 | 3 | −3 | $k^{-4}$ | $r^{2}$ | $k^{-5}$ | 1 | 3.0 | 3 | −3 | $\frac{4}{3}$ |

Key:

| | |
|---|---|
| Real | 2 point correlation function in real space |
| FT | Spectral density function = Fourier transform of correlation function |
| r | Lag distance in real space |
| k | Wavenumber in Fourier space |
| H | Hurst exponent of fluctuations (some values are nominal only) |
| D | Fractal dimension of fluctuations (some values are nominal only) |
| β | Exponent of wavenumber in spectral density function $\sim k^{-\beta}$ |
| $\eta$ | Critical exponent of correlation function $\sim r^{-(d-2+\eta)}$ |
| $\omega_c$ | Critical value of normalization constant ($\eta = d + 2 - 2d\omega_c$) |
| fBm | Fractional Brownian motion model |
| * | Landau-Ginzburg model has similar exponents to those of mean-field theory, except that the best value of $\eta \sim 0.032 \mp 0.005$ |

strain is a satisfactory, if approximate, order parameter for renormalization. Then, because strain averages arithmetically, it is necessarily associated with $\omega_c \sim 1$ and $\eta \sim -1$. It is worth noting that the 'high temperature' value of $\omega = ½$ , corresponding to $\eta = +2$, would imply $k^{+2}$ scaling of strain in one-dimensional samples, or $k^0$ scaling of strain in three-dimensions; i.e., elastic behavior. Table 2 summarizes the various models in various dimensions of strain fluctuations, and their associated scaling exponents.

## *Correlations of Induced Seismicity*

The seismic slip displacements induced by water injection into a well in a granite geothermal site (STARZEC *et al.*, 2000) have also been investigated for spatial correlations. The variograms in Figure 4 of STARZEC *et al.* (2000) are of individual

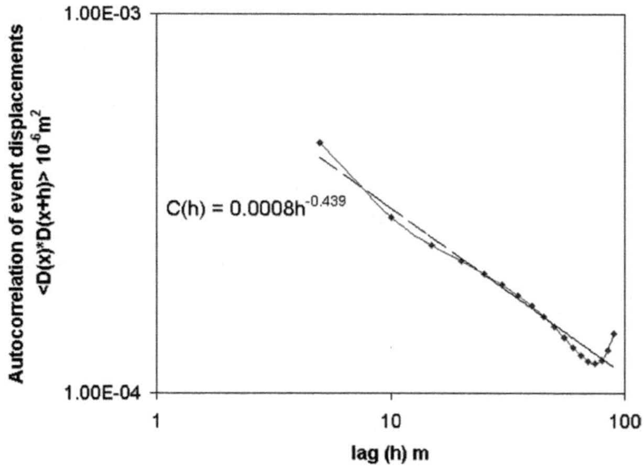


Figure 4
Autocorrelation of seismic slip displacements interpreted from source parameters of induced earthquakes during water injection at the Soultz-sous-Fôrets HDR site. Data aggregated over full injection interval from graphs in STARZEC *et al.* (2000). Log-log plot suggests a power-law with exponent ~−0.44.

slips on fractures, which approximate increments in local strain. The graphs of Figure 4 each pertain to a limited depth interval; the authors point out that the variograms are sensible or not depending upon whether there is just one, or multiple, fracture sets within the interval. Aggregation of the data from the intervals should reduce the orientational dependency and emphasize the dependence on lateral lag distance.

Figure 4 of this paper presents the autocorrelation function of seismic slip displacements aggregated from the total interval of microseismicity. The fitted power-law function $C(h) \sim h^{-0.44}$ is consistent with the gentle decline of strain correlation with distance predicted above from scale-dependent effective moduli.

## *Discussion*

Table 3 summarizes the scaling exponents predicted by renormalization and those inferred from empirical data above. It is seen that although the empirical data support high negative values of the scaling exponent of effective modulus, there is nevertheless a significant range of values. There are several potential reasons for this range, depending upon the type of data. The possible influences of erosional or sedimentary morphologies on depths to a surface have already been mentioned. The possibility that fluctuations in well-log values are not solely derived from strain variations is clear. Furthermore, both surface displacements and relict strain fields are the cumulative results of deformational processes that have proceeded through

Table 3

*Summary of scaling exponents predicted by renormalization theory and inferred from empirical data. Refer to text for descriptions of each case*

| Case | Dimension of process | Scaling exponent of wavenumber ($k$) in 3D spectral density of: | | Equivalent scaling exponent of effective modulus in terms of: | |
|---|---|---|---|---|---|
| | | Strain | Displacement | Wavenumber, $k$ | Domain size, $r$ |
| Renormalization approach of this paper | 2-D | −2.5 | −4.5 | +2.5 | −2.5 |
| | 3-D | −4 | −6 (min for correlation fn) | +4.2 | −4.2 |
| Numerical random bond network near percolation | 2-D | | | | −3.0 |
| | 3-D | | | | −4.3 |
| Depths to seismic horizon | 3-D | N/A | −4.5 to −5.0 | +2.5 to 3.0 | −2.5 to −3.0 |
| '$1/k$' heterogeneities | 3-D | −3.0 to −3.6 | −5.0 to −5.6 | 3.0 to 3.6 | −3.0 to −3.6 |
| Microseismicity induced by water injection | 2-D ? | ($r^{-0.44}$) −(3.0-0.44) = −2.56 | −4.56 | 2.56 | −2.56 |

multiple stages. Although the deformational state might have been close to criticality for most of such stages, the statistics of resultant deformational parameters can only be expected to approximate those suited to a persistent critical state. The slip displacements due to water injection might be closer to representing instantaneous deformation in a persistent state of criticality, but the anisotropy in the clouds of induced seismicity described by STARZEC *et al.* (2000) imply that the dimensionality of the process might be intermediate between 2 and 3, rather than purely three-dimensional. Further work on both theoretical and empirical aspects of the above is required to narrow the range of scaling exponents in a particular application.

## *Conclusion*

A real-space renormalization scheme has been devised that allows the exponent of the spatial scaling of effective modulus near the critical point of fracturing to be determined. The most appropriate resultant negative exponents are ~2.5 and 4.2 in two- and three-dimensional space, respectively. The high negative values are approximately consistent with a) previous numerical calculations of scaling of modulus on random bond networks near the percolation density; b) empirical observations of depth profiles of surfaces; c) '$1/k$' scaling of fluctuations in one-dimensional samples of geological heterogeneities; d) spatial correlation of micro-seismic emissions resulting from water injection. The power-law spatial scaling of effective modulus has been substituted for the modulus appearing in the directional form of the 2-point correlation of elastic displacements. Although deformational statistics might be more closely characterized by multifractality, the resultant correlation function has been used to interpolate the full displacement vector across a field conditioned on its upper and lower surfaces to give a map of maximum principal strain that shows more realism as the scaling becomes more long-range.

*Acknowledgements*

Part of the work leading to this paper was financed by BP Exploration Ltd., the Department of Trade and Industry (UK), and Roxar Ltd. Very helpful discussions were also held with Dr. Colin Daly at Roxar Ltd, Professor Stuart Crampin at the University of Edinburgh and Dr. Peter Leary.

*Appendix: Spatial Correlation of Fractures, Elastic Case.*

The following is an adaptation of Chapter 4 of LANDAU and LIFSHITZ (1975) on the treatment of elastic deformations in the presence of a distribution of dislocations.

The total energy per unit volume is the work done by the elastic forces on the total strain. That is $E = \frac{1}{2}\sigma^e_{ij}e^e_{ji}$. All terms in this have well-defined Fourier Transforms so we can work throughout in Fourier space. The definition of elastic strain is

$$e^e_{ij} = \left(\tfrac{i}{2}\right)\left(u^e_i k_j + u^e_j k_i\right) = \left(\tfrac{i}{2}\right)\left(\delta_{aj}\delta_{il} + \delta_{ai}\delta_{jl}\right)k_a u^e_l.$$

Similarly the elastic contribution to the stress can be written $\sigma^e_{ij} = i\lambda_{ijal}k_a u^e_l$, where $\lambda_{ijal}$ is the standard isotropic elasticity tensor, related to the Lamé constants, $\lambda$ and $\mu$, by $\lambda\delta_{ij}\delta_{al} + \mu\left(\delta_{ia}\delta_{jl} + \delta_{il}\delta_{ja}\right)$. Contracting these terms together leads to an expression for the elastic energy:

$$E = \tfrac{1}{2}\left[(\lambda+\mu)k_k k_l + \mu k^2\delta_{kl}\right]u^e_k(\underline{k})\,u^e_l(-\underline{\mathbf{k}}) = \tfrac{1}{2}\mu\left[k^2\delta_{kl} + \left(\tfrac{1}{1-2\upsilon}\right)k_k k_l\right]u^e_k(\underline{\mathbf{k}})\,u^e_l(-\underline{\mathbf{k}})$$

$$= \tfrac{\mu}{2}L_{kl}(\underline{\mathbf{k}})\,u^e_k(\underline{\mathbf{k}})\,u^e_l(-\underline{\mathbf{k}}),$$

where $L_{ij}$ is the usual linear operator of isotropic elasticity and is the inverse of the Green's function, $L_{ij}G_{jk} = \delta_{ik}$.

The Boltzmann distribution for the elastic energy can be expressed as

$$p(E) = Z^{-1} \exp\left(-\frac{1}{\langle E\rangle}\int d^3k\, E(k)\right) = Z^{-1} \exp\left(-\frac{\mu}{\langle E\rangle}\int d^3k\, L_{kl}(\underline{\mathbf{k}})\, u_k^e(\underline{\mathbf{k}})\, u_l^e(-\underline{\mathbf{k}})\right)$$

where $\langle E\rangle$ is the mean energy and $Z$ is the partition function (normalization constant). Assuming Boltzmann statistics, it is then easy to demonstrate that the spatial correlation in the elastic displacement is given by:

$$c_{ij}(\underline{\mathbf{k}}) = \left\langle u_i^e(\underline{\mathbf{k}})\, u_j^e(-\underline{\mathbf{k}})\right\rangle = \frac{\langle E\rangle}{2\mu} L_{ij}^{-1} = \frac{\langle E\rangle}{2\mu} G_{ij} = \frac{\langle E\rangle}{2\mu k^2}\left[\delta_{ij} - \frac{1}{2(1-\upsilon)}\frac{k_i k_j}{k^2}\right].$$

In real space the correlation of displacements takes the form:

$$\left\langle u_i^e(\underline{\mathbf{r}}).u_j^e(\underline{\mathbf{0}})\right\rangle = \frac{\langle E\rangle}{16\pi(1-\upsilon)}\frac{1}{r}\left[(3-4\upsilon)\delta_{ij} + \frac{r_i r_j}{r^2}\right].$$

There is some evidence that the energy actually follows a power-law (i.e., $P(E) \sim E^{\alpha}$), attributable to the degeneracy of states with a given energy that dominates the Boltzmann distribution near the critical point (MAIN and BURTON, 1994; MAIN and AL-KINDY, 2002). Note that a power-law distribution is not normalizable unless a cut-off is applied (whether or not this is an ultra-violet or an infra-red cut-off depends on the value of $\alpha$). In fact the spatial correlation is very similar to that for the Boltzmann distribution in that the correlations essentially depend on the inverse of the kernel of the energy function. In addition there are some multiplicative terms involving the cut-off which is required to ensure that the integrals are well defined. It is readily shown that the essential correlation structure is the same as that for the Boltzmann distribution.

## REFERENCES


ALLÈGRE, C.J., LE MOUËL, J.L., and PROVOST, A. (1982), *Scaling rules in rock fracture and possible implications for earthquake prediction*, Nature *297*, 47–49.

ALLÈGRE, C.J. and LE MOUËL J.L. (1994), *Introduction of scaling techniques in brittle fracture of rocks*, Phys. Earth Planet Inter. *87*, 85–93.

ARBABI, S. and SAHIMI, M. (1988), *Absence of universality in percolation models of disordered elastic media with central forces,* J. Phys. A: Math. Gen. *21*, L863–L868.

BAK, P., *How Nature Works – the Science of Self-Organized Criticality* (Oxford University Press, Oxford 1997). BINNEY, J.J., DOWRICK, N.J., FISHER, A.J., and NEWMAN, M.E.J. *The Theory of Critical Phenomena – An Introduction to the Renormalization Group* (Oxford University Press, Oxford 1992).

BRUNER, W.M. (1976), *Comment on 'Seismic velocities in dry and saturated cracked solids' by O'Connell and Budiansky*, J. Geophys. Res. *81*, 2573–2576.

BUDIANSKY, B. and O'CONNELL, R.J. (1976), *Elastic moduli of a cracked solid*, Int. J. Solids Struct. *12*, 81–97.

CHAKRABARTI, B.K. and BENGUIGUI, L.G., *Statistical Physics of Fracture and Breakdown in Disordered Systems* (Oxford University Press, Oxford 1997).

COWIE, P. A., SORNETTE, D., and VANNESTE C. (1995), *Multifractal scaling properties of a growing fault population*, Geophys. J. Int. *122*, 457–469.

FENG, S., *Elasticity and percolation* in *scaling phenomena.* In *Disordered Systems* (eds. R. Pynn and A. Skjeltorp), (NATO ASI Series B: Physics, vol.133, Plenum Press, New York and London 1985).

FENG, S. and SEN, P.N. (1984), *Percolation on elastic networks: New exponent and threshold,* Phys. Rev. Lett. *52,* 216.

GUYON, E., HANSEN, A., HINRISCHEN, E., and ROUX, S. (1989), *Critical behaviors of central-force lattices,* Physica A *157*, 580–586.

HEFFER, K.J., KING, P.R., and JONES, A.D.W. (1999), *Fracture modelling as part of integrated reservoir characterization,* Society of Petroleum Engineers paper SPE 53347.

HENYEY, F.S. and POMPHREY, N. (1982), *Self-consistent moduli of cracked solid,* Geophys. Res. Lett. *9*, 903–906.

HUANG, Y., CHANDRA, A., JIANG, Z.Q., WEI, X., and HU, K.X. (1996), *The numerical calculation of two-dimensional effective moduli for microcracked solids,* J. Solids Struc. *33,* 1575–1586.

HUDSON, J. A. (1980), *Overall properties of a cracked solid,* Math. Proc. Cambridge Phil. Soc. *88*, 371–384.

HUDSON, J.A. (1981), *Wave speeds and attenuation of elastic waves in material containing cracks.* Geophys. J. Roy. Astr. Soc. *64*, 133–150.

HUDSON, J.A. (1986), *A higher-order approximation to the wave propagation constants for a cracked solid,* Geophys. J. Roy. Astr. Soc. *87*, 265–274.

JAEGER, J.C. and COOK, N.G.W., *Fundamentals of Rock Mechanics, Third Edition* (Chapman and Hall, London, 1979).

KADANOFF, L.P. (1966), *The introduction of the idea that exponents could be derived from real-space scaling arguments*, Physics *2*, 263–273.

KALLESTAD, E. (1998), *Stochastic simulation of Gaussian vector fields*, Diploma Thesis in the Faculty of Physics, Informatics and Mathematics, Norwegian University of Science and Technology, (NTNU) Trondheim.

KANTOR, Y. (1985), *Elastic properties of random systems.* In *Scaling Phenomena in Disordered Systems* (eds. R. Pynn and A. Skjeltorp), (NATO ASI Series B: Physics, vol. 133, Plenum Press, New York and London 1985).

KELVIN, LORD (William Thomson) (1856), *Elements of a mathematical theory of elasticity, Part 1, On stresses and strains*, Phil. Trans. Roy. Soc. *166*, 481–498.

KEMENY, J. and COOK, N.G.W. (1986), *Effective moduli, non-linear deformation and strength of a cracked, elastic solid,* Int. J. Rock Mech. Min. Sci. and Geomech. Abstr. *23*, 2, 107–118.

KING, P.R. (1989), *The use of renormalization for calculating effective permeability*, Trans. Por. Media *4*, 37–58.

LANDAU, L. D. and LIFSHITZ, E. M., *Theory of Elasticity* (Pergamon, New York 1975).

LEARY, P.C., *Fractures and physical heterogeneity in crustal rock.* In *Heterogeneity in the Crust and Upper Mantle: Nature, Scaling and Seismic Properties* (eds. Goff, J.A. and Holliger, K.) (Kluwer Academic, New York 2002).

LEMIEUX, M.A., BRETON, P., and TREMBLAY, A-M.S. (1985), J. Phys. (Paris) Lett. *46,* L-1.

MAIN, I.G. (1996), *Statistical physics, seismogenesis and seismic hazard*, Rev. Geophys. *34*, 4, 433–462.

MAIN, I.G. and AL-KINDY, F.H. (2002), *Entropy, energy and proximity to criticality in global earthquake populations*, Geophys. Res. Lett. *29*, 7, 25–1:4.

MAIN, I.G. and BURTON, P.W. (1984), *Information theory and the earthquake magnitude frequency distribution*, Bull. Seismol. Soc. Am. *74*, 1409–1426.

MARSAN, D. and BEAN, C.J. (2003), *Multifractal modelling and analyses of crustal heterogeneity.* In *Heterogeneity in the Crust and Upper Mantle: Nature, Scaling and Seismic Properties* (eds. Goff, J.A. and Holliger, K.) (Kluwer Academic, New York 2002).

MEHRABADI, M.M. and COWIN, S.C. (1990), *Eigentensors of linear anisotropic elastic materials*, Mech. Appl. Math. *43*, Pt 1, 15–41.

O'CONNELL, R.J. and BUDIANSKY, B. (1974), *Seismic velocities in dry and saturated cracked solids,* J Geophys. Res. *79*, 5412–5426.

OUILLON G., CASTAING, C., and SORNETTE, D. (1996), *Hierarchical geometry of faulting*, J. Geophys. Res. *191*, 5477–5487.

RODIN, G.J. and HWANG, Y.L. (1991), *On the problem of linear elasticity for an infinite region containing a finite number of non-intersecting spherical inhomogeneities,* Int. J. Sol. Struct. *27,* 145–159.

SALGANIK, R.L. (1973), *Mechanics of bodies with many cracks*, Mech. Solids *8*, 135–143.

SAYERS, C. and KACHANOV, M. (1991), *A simple technique for finding effective elastic constants of cracked solids for arbitrary crack orientation statistics*, Int. J. Sol. Struct. *27*, 6, 671–680.

SHEN, L. and LI, J. (2004), *A numerical simulation for effective elastic moduli of plates with various distributions and sizes of cracks*, Int. J. Sol. Struct. *41*, 7471–7492.

SHEN, L. and YI, S. (2001), *An effective inclusion model for effective moduli of heterogeneous materials with ellipsoidal inhomogeneities,* Int. J. Sol. Struct. *38,* 5789–5805.

SMALLEY, J., TURCOTTE, D.L., and SOLLA, R.J. (1985), *A renormalization group approach to the stick-slip behaviour of fault,* J. Geophys. Res. *90,* 1894–1900.

STARZEC, P., FEHLER, M., BARIA, R., and NIITSUMA, H. (2000), *Spatial correlation of seismic slip at the HDR-Soultz geothermal site: Qualitative approach*, Bull. Seismol. Soc. Am. *90*, 6, 1528–1534.

TURCOTTE, D.L. (1986), *Fractals and fragmentation,* J. Geophys. Res. *91,* 1921–1926.

WANG, J., FANG, J., and KARIHALOO, B.L. (2000), *Asymptotic bounds on overall moduli of cracked bodies,* Int. J. Sol. Struct. *37,* 6221–6237.

ZIMMERMAN, R.W. (1985), *The effect of microcracks on the elastic moduli of brittle materials*, J. Mater. Sci. Lett. *4*, 1457–60.

ZIMMERMAN, R.W., *Compressibility of Sandstones. Developments in Petroleum Science*. (Elsevier, Amsterdam 1991a).

ZIMMERMAN, R.W. (1991b), *Elastic moduli of a solid containing spherical inclusions,* Mech. Mater. *12,* 17–24.

ZHOU, M. and SHENG, P. (1993), *Shear rigidity in 2D solid-liquid composites,* Phys. Rev. Lett. *71,* 26, 4358–4360.

(Received March 15, 2005, revised December 1, 2005, accepted January 4, 2006)

Pure appl. geophys. 163 (2006) 2243–2258
0033–4553/06/102243–16
DOI 10.1007/s00024-006-0116-0



Pure and Applied Geophysics

# Analysis of Temporal Structures of Seismic Events on Different Scale Levels

V. I. GERMAN

*Abstract*—A statistical model for describing the energy scaling of the distribution of inter-event times is described. By considering the diverse region seismicity (natural and induced) on different scale (energy/magnitude) levels the self-similarity of the distribution has been determined. A comparison between the distribution of inter-event times on different scale levels and the most popular distributions of reliability theory has been carried out. The distribution of inter-event times for different scale levels is well approximated by the Weibull distribution. The Weibull distribution, with parameters which obey the scaling model and the Gutenberg-Richter law, has been tested.

**Key words:** Waiting times, seismicity, scaling of distributions.

## *1. Introduction*

The temporal structure of seismicity (registered sequence of seismic events) together with the spatial and energy structures are general characteristics required for understanding of physical processes in the earth's crust and for seismic hazard assessment (e.g., SOBOLEV *et al.*, 1991; TOMILIN and VOINOV, 1995; KNOPOFF *et al.*, 1996; ULOMOV *et al.*, 1999).

The investigation of inter-event times between successive earthquakes has a long history, but in most cases it concerns only seismicity in fixed regions without variation of energy-time-space domain of analysis and thus without checking its self-similarity properties (e.g., GARDNER and KNOPOFF, 1974; RIKITAKE, 1976; NISHENKO and BULAND, 1987; MUKHAMEDOV, 1996; CORREIG *et al.*, 1997; WANG and KUO, 1998) or only with a graphical analysis of obtained distributions, without application statistical tests (BAK *et al.*, 2002; GERMAN, 2002; CORRAL, 2003, 2004; DAVIDSEN and GOLTZ, 2004).

Recent research considers the inter-event times intervals as the stochastic variable and investigates its probability density function with variation of energy-time-area

Siberian Aerospace State University named after M. F. Reshetnev, 31 Krasnoyarsky rabochy st., 660014, Krasnoyarsk, Russia. E-mail: germanv@rambler.ru domain (BAK *et al.*, 2002; GERMAN, 2002, 2005; CORRAL, 2003, 2004; DAVIDSEN and GOLTZ, 2004). However, arbitrary variation of this domain has shown that the shape of distribution of inter-event times for them is not quite universal (DAVIDSEN and GOLTZ, 2004). Some theoretical questions which concerned inter-event times intervals distribution are discussed in Sornette and Knopoff (1997) and MOLCHAN (2005).

At present, a very popular idea exists about the self-similarity of the seismic/failure process in the earth's crust/rocks on different scale levels: from the micro-level to strong earthquakes (SADOVSKY *et al.*, 1987; RYKUNOV *et al.*, 1987; TURCOTTE, 1992; KUKSENKO *et al.*, 1996; MUKHAMEDOV, 1996; PONOMAREV *et al.*, 1997; ULOMOV, 1998; GERMAN and MANSUROV, 2002; BAIESI and PACZUSKI, 2004).

The most famous and detailed investigated relationship of self-similarity in seismology is the Gutenberg-Richter (G-R) law (e.g., KAGAN, 1999), which shows the self-similarity in energy radiation for different scale (energy/magnitude) levels. For a fixed space area, the temporal structure is determined by the set of the inter-event times for each scale level of seismicity (IETSL), and the average value of IETSL is inversely proportional to the number of events for each level. Thus the G-R law can be considered as the scale law for the average value of IETSL. Moreover it can be used to scale the distribution of inter-event times (see below).

The present paper deals with the following question: is the temporal structure of seismicity (distributions of temporal) stable with the variation of energy (magnitude) ranges only or not? To answer this question the statistical accelerated life model is considered and tested. Advanced statistical analysis helps to obtain the distribution type of IETSL and to determine the physical features of seismicity on different scale levels. The final investigations are devoted to the relation between the distribution type, the model and the G-R law to confirm the consistency of the results obtained.

## *2. The Properties of Inter-event Times for Different Scale Levels of Seismicity*

The concept of the self-similarity of the seismic process on different scale levels also includes a temporal self-similarity. According to this conception seismicity on one-scale level is a scaled version of seismicity on another level. This similarity is not absolute, but statistical. Let us consider seismic inter-event times for each energy-time-area interval as the stochastic variable $\Delta T$, then

$$\Delta T = \Delta T_{\varepsilon} = \Delta T_{\varepsilon_0} / \varphi(\vec{\varepsilon}), \tag{1}$$

(GERMAN, 2002) where $\varphi(\vec{\varepsilon})$ is a scale coefficient, which is a function of parameters of the energy-time-area interval, which is defined by the vector of its parameters $\vec{\varepsilon}$.

This corresponds to the statistical accelerated life model (COX and OAKES, 1984). According to the model, the lifetime of an object at a scale level is obtained as a pressed or unpressed version of the life time of the object, on base level with $\vec{\varepsilon} = \vec{\varepsilon}_0$.

Thus, according to this model (COX and OAKES, 1984) the inter-event times distribution (cumulative distribution) function of the variable $\Delta T$ will obey the relationship

$$F(\Delta t, \vec{\varepsilon}) = F[\Delta t \varphi(\vec{\varepsilon}), \vec{\varepsilon}_0] = F_0(\Delta t \varphi(\vec{\varepsilon})) \tag{2}$$

and the probability density function, $f(\Delta t, \vec{\varepsilon})$, will obey

$$f(\Delta t, \vec{\varepsilon}) = \frac{dF(\Delta t, \vec{\varepsilon})}{d(\Delta t)} = \varphi(\vec{\varepsilon}) f[\Delta t \varphi(\vec{\varepsilon}), \vec{\varepsilon}_0] = \varphi(\vec{\varepsilon}) f_0(\Delta t \varphi(\vec{\varepsilon})),$$

where $F_0$ and $f_0$ are constant functions, the variable $\Delta t$ corresponds to stochastic variable $\Delta T$.

To determine the scale parameter of the model, it is necessary to calculate a mean for both parts of Eq. (1). Thus, the general form for the scale coefficient is

$$\varphi(\vec{\varepsilon}) = M(\Delta T_{\varepsilon 0})/M(\Delta T) \cong M(\Delta T_{\varepsilon 0})(N_\varepsilon/T) = \text{const.}(N_\varepsilon/T), \tag{3}$$

where $T$ is the duration of the considered time period and $N_\varepsilon$ is the number of events for the energy-time-area interval considered (here we take into account that the average value of inter-event times is a good estimation of its mean).

To obtain a model for inter-event times for the different energy-time-area intervals, it is necessary to find the logarithm of Eq. (1) and to carry out the substitution $\log_{10} \Delta T_{\varepsilon 0} = M(\log_{10} \Delta T_{\varepsilon 0}) + \theta$ (COX and OAKES, 1984):

$$\log_{10} \Delta T = \rho_0 - \log_{10} \varphi(\vec{\varepsilon}) + \theta, \tag{4}$$

where $\rho_0 = M(\log_{10} \Delta T_{\varepsilon 0})$, $\theta$ is the stochastic variable with a mean equal to zero and a distribution function which does not depend on $\vec{\varepsilon}$ (COX and OAKES, 1984). According to this model, the distributions of $\log_{10} \Delta T$ for different values of $\vec{\varepsilon}$ are the same, but have shifted, so that the standard deviation is constant for all energy-time-area intervals

$$\sigma(\log_{10} \Delta T) = \text{const.} \tag{5}$$

The mean for both parts of Eq. (3)

$$M(\log_{10} \Delta T) = -\log_{10} \varphi(\vec{\varepsilon}) + \rho_0. \tag{6}$$

Equations (5) and (6) are valid if the general assumption (1) regarding the united scaling law for inter-event times is correct. However, it seems impossible to avoid a significant deviation between scaled distributions of inter-event times for all values of $\vec{\varepsilon}$ (for arbitrary energy-time-area intervals) (BAK *et al.*, 2002; CORRAL, 2004; DAVIDSEN and GOLTZ, 2004).

Thus, in this paper, to obtain solid results about self-similarity of inter-event times, only the energy scaling has been considered (time-space interval is fixed and $\vec{\varepsilon} = E$). To calibrate the accelerated life model for each region, the classical G-R law

$$\log_{10}(N_E/T) = -\gamma \log_{10} E + \log_{10} d_E$$

has been used, where $N_E$ is the number of events, with the logarithm of radiated energy given by $(\log_{10} E(J) - 0.5;\ \log_{10} E(J) + 0.5)$. In this case, according to (3) and (6) $\varphi(\vec{\varepsilon}) = \varphi(E) = d_E M(\Delta T_{\varepsilon_0})/E^{\gamma} = \text{const.}/E^{\gamma}$ and

$$M(\log_{10} \Delta T) = \gamma \log_{10} E + \delta_0, \tag{7}$$

where $\delta_0 = \rho_0 - \log_{10}(d_E M(\Delta T_{\varepsilon_0}))$ and $\Delta T$ is IETSL for each region. Thus the mean of IETSL logarithm $M(\log_{10} \Delta T)$ has a linear relationship with the energy class $\log_{10} E(J)$. The constant value of the standard deviation $\sigma(\log_{10} \Delta T)$ allows the classical form of the least-squares method to be utilized to estimate the parameters $\gamma$ and $\delta_0$ effectively. This is an advantage of this model over the G-R law. Another is that, according to the accelerated life model, it is possible to not only estimate the average number of strong seismic events, but also the distribution of inter-event times for them, by analyzing the temporal structure of the weak events.

It is also possible to apply the accelerated life model for different types of energy, time and space scaling of distributions of the inter-event times and for scaling of the distribution of other physical values, such as, for example, the distances between seismic events (GERMAN, 2005).

## *3. Experimental Data*

Experimental checking of the inferences concerning the properties of IETSL structure has been carried out for two types of seismic events, induced and natural, over a wide range of scales of seismic process, covering ~14 orders of magnitude of radiated energy or 14 energy classes (Table 1). For the analysis of natural seismicity, the *Special Catalogue of Earthquakes of Northern Eurasia* (to calculate the energy class from local magnitude, the formula $\log_{10} E(J) = 1.8M_L + 4.0$, RAUTIAN, 1964 was used) and a catalogue of earthquakes in the Toktogul region ($M_L = (\log_{10} E(J) - 1.3)/3$ for $M_L < 1.8$, $M_L = (\log_{10} E(J) - 4)/1.8$ for $M_L \geq 1.8$), created at the United Institute of the Physics of the Earth of the Russian Academy of Sciences have been used. The catalogues and additional descriptions of them are available on the Internet (see references list).

Induced seismicity is represented by catalogues of the North Ural Bauxite Mine and Gold fields Welkom (ISS International, South Africa). The average seismic activity for these catalogues does not vary greatly in the period considered. The daily

Table 1

*Characteristics of regions considered*

| Region | Induced seismicity | | Natural seismicity: Seismoactive regions | | | |
|---|---|---|---|---|---|---|
| | North Ural Bauxite Mine | Gold fields Welkom (RSA) | Toktogul | Baikal | Kamchatka | Kamchatka-Kurils |
| Analyzed period, years | 1984... 1989 | 01.01.1995... 31.08.1995 | 1965... 1991 | 1962... 1990 | 1962... 1990 | 1962...1990 |
| Location | X: 6 400... 7 130 m<br>Y:1 600... 1 940 m<br>Z: −110... −980 m | X: 900...19 600 m<br>Y: 350... 15 000 m<br>Z: −490... −3 920 m | 39.8... 42.7° N.lat.<br>69.9... 74.5° E.lon. | 51... 60.6° N.lat.<br>100...120° E.lon. | 51... 58° N.lat.<br>154...165° E.lon. | Area with apexes, (°N.lat.;° E.lon.): (42; 140), (42; 153), (57; 167), (57; 158), (48; 151), (43; 140) |
| Number of events | 1 200 | 5 400 | 8 800 | 1 700 | 6 100 | 9 500 |
| Energy class of events | 1.6...6.9 | 1.3...10.3 | 1.3... 13.9 | 10.3... 14.4 | 10.3... 18.4 | 10.3...18.8 |
| Energy class completeness threshold $\log_{10} E_{\min}(J)$ | 3.3* | 5.4 | 8.0 | 10.8 | 10.8 | 13.0 |
| Interval obeys the accelerated life model | 3.3...5.7 | 5.4...8.5 | 8.0... 0.3 | 10.8...14.4 | 10.8... 15.2 | 13.0...16.7 |

* Here and further the central value of energy class interval is written.

variability of the mining activity on the North Ural Bauxite Mine influenced the inter-event times distribution up to energy class $\log_{10} E(J) = 3.5$. For the North Ural Bauxite Mine (GERMAN, 2002; GERMAN and MANSUROV, 2002; TOMILIN and VOINOV, 1995), the local area, with high activity and numerous strong rockbursts was considered. The large areas were considered for earthquakes (Table 1).

In each region, the G-R plots were used to determine the energy class completeness threshold, $\log_{10} E_{\min}(J)$, above which a catalogue can be considered to be reasonably complete. This corresponds to the point where the G-R plot significantly changes its slope. The considerable data in the catalogues allow a detailed analysis of the IETSL structure to be carried out. Further important relationships are demonstrated with examples from the North Ural Bauxite Mine and the Kamchatka-Kurils seismoactive region, as extreme scale level representatives.

## 4. *Checking of Inter-event Times Self-similarity on Different Scale Levels*

In Section 2 the relationships (5) and (7) for the mean and for the standard deviation of logarithm of seismic inter-event times on different scale levels of seismicity are obtained. They allow the experimental checking of its self-similarity to be carried out. The correspondent relations are shown in Figure 1. Each point on this figure was obtained for seismic events in the energy interval $(E/10^{0.5};\ 10^{0.5}E)$; that is, from one energy class $(\log_{10} E(J) - 0.5;\ \log_{10} E(J) + 0.5)$. This interval size allows a sufficient number of data to be selected out for the statistically valid analysis and to control the local features of IETSL structure. A step of the shift for this interval is 0.2; only for the North Ural Bauxite Mine it is

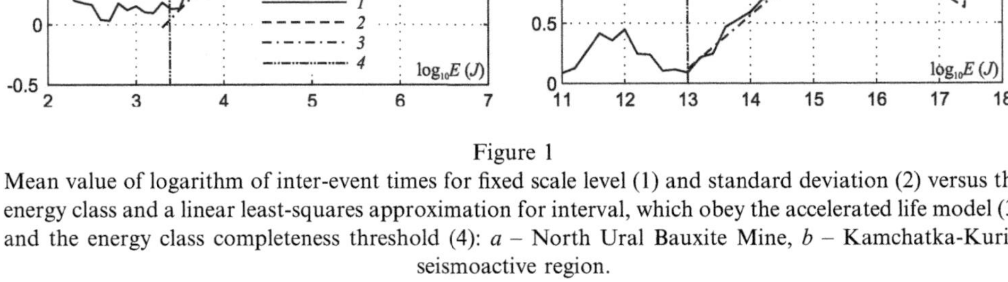


Figure 1
Mean value of logarithm of inter-event times for fixed scale level (1) and standard deviation (2) versus the energy class and a linear least-squares approximation for interval, which obey the accelerated life model (3) and the energy class completeness threshold (4): *a* – North Ural Bauxite Mine, *b* – Kamchatka-Kurils seismoactive region.

0.1 because of the small number of seismic events. The same division into scale levels will be used for the further analysis.

Intervals corresponding to the accelerated life model (i.e., they obey equations (5) and (7)) exist for all considered regions; they begin at the energy class completeness threshold $\log_{10} E_{\min}(J)$ (see Table 1). The linear part of the $M(\log_{10} \Delta T)$ plot continues to nearly the maximum energy classes, except at the North Ural Bauxite Mine, where the strongest rockbursts, with $\log_{10} E(J) > 6.5$, do not have a clearly induced nature, but a mixed induced-tectonic nature (TOMILIN and VOINOV, 1995). This feature leads to a bimodal (two linear intervals) type of G-R plot. At the same time, $\sigma(\log_{10} \Delta T)$ in areas of strong events often has a significant deviation from a constant value. The existence of intervals which obey the accelerated life model for all regions allows the parameters $\gamma$ and $\delta_0$ in (7) to be used as seismic regime characteristics for each of the regions as the parameters of the G-R law are used. The relationships (5) and (7) can also be used to determine the energy class completeness threshold.

The behavior of the curves in Figure 1a, for the interval $\log_{10} E(J) \in (3.3; 5.7)$, corresponds to the model described and hence the seismic process at different scale levels for this interval is self-similar. This energy interval correlates well with estimates obtained by German and Mansurov (GERMAN and MANSUROV, 2002). A linear least-squares approximation of the relationship is $M(\log_{10} \Delta T(\text{days})) = 0.68 \log_{10} E(J) - 2.24$. The same circumstance for the interval $\log_{10} E(J) \in (13.0; 16.7)$ is given in Figure 1b: $M(\log_{10} \Delta T(\text{days})) = 0.45 \log_{10} E(J) - 5.66$.

## *5. Investigation of Inter-event Times Distribution Type*

The next step in the IETSL investigation was to determine its distribution. Essentially a distribution must be determined which describes all the experimental data, obeys both the G-R law and the accelerated life model and further allows the interaction between events to be described. The last point, which mathematically corresponds to the aftereffect of events arising, is very important, because the interaction leads the seismic process from one scale level to the next, higher one (see KUKSENKO *et al.*, 1996; GERMAN and MANSUROV, 2002).

In many papers, the question about inter-event times distribution is considered, but in most cases without division into different scale levels and thus without checking its self-similarity properties. The most popular distribution for inter-event times is the exponential distribution. This corresponds to a simple stream (stationary Poisson process), however in most cases the approximation of the experimental data to an exponential distribution ($F(\Delta t) = 1 - \exp(-\lambda_0 \Delta t)$) is not sufficiently good. At the same time, some researchers have shown that for some data sets it is possible to obtain a good fit by applying the lognormal $\left(f(\Delta t) = \frac{1}{\tau\sqrt{2\pi}} \exp\left(-\frac{(\ln(\Delta t) - \zeta)^2}{\tau^2}\right)\right)$, gamma ($f(\Delta t) = \lambda_\Gamma(\lambda_\Gamma \Delta t)^{k_\Gamma - 1} \exp(-\lambda_\Gamma \Delta t)/\Gamma(k_\Gamma)$, $\Gamma(x)$ – gamma function) or Weibull ($F(\Delta t) = 1 - \exp(-\lambda \Delta t^k)$) distribution (e.g., RIKITAKE, 1976; NISHENKO and BULAND, 1987; MUKHAMEDOV, 1996; CORREIG *et al.*, 1997; CORRAL, 2004), which have the similar shape in this case.

To estimate the type of IETSL distribution, the dimensionless relationship between the normalized third central moment, $\mu_3 = M[(\log_{10} \Delta T)^3]/\sigma^3(\log_{10} \Delta T)$, and the coefficient of variation, $\chi = \sigma(\log_{10} \Delta T)/M(\log_{10} \Delta T)$, for all regions was constructed for scale levels larger than the energy class completeness threshold, $\log_{10} E_{\min}(J)$ (Table 1). These were then compared with the theoretical curves for the most popular families of distributions of the reliability theory (COX and OAKES, 1984). In such a coordinate system, each of these families of distributions, for all values of the parameters, is represented as a single curve. This allows a very quick and easy comparison of the experimental data with many families of distributions simultaneously to be made and to obtain an initial estimation of their distribution type. This method should be considered as a variant of a statistical method of moments and hence not as a precise method.

It was determined that the experimental points for all regions are close to the curve which corresponds to the Weibull family of distributions and near the curve of the gamma family of distributions (both these families include the exponential family of distributions) (Fig. 2).

The points in Figure 2 and in figures for other regions can be divided into two groups: The minimum deviation from the theoretical curve is for the energy intervals, which correspond to weak events, with a coefficient of variation generally $> 1.3$. The other points correspond to the intervals between relatively strong events;

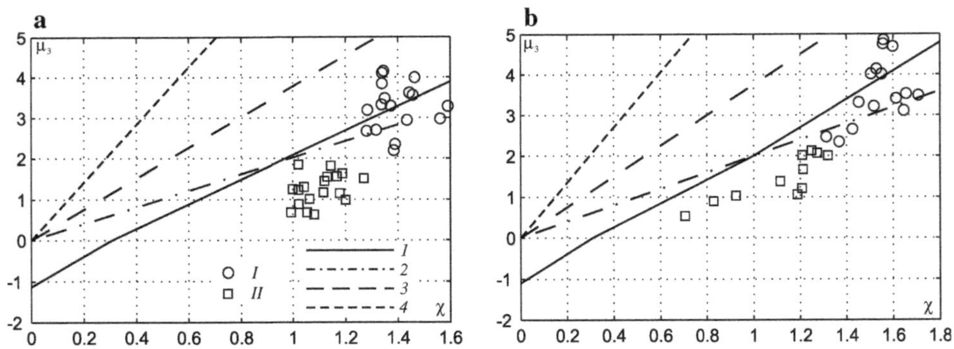


Figure 2
Comparison of the experimental points with theoretical curves (normalized central moment $\mu_3$ versus coefficient of variation $\chi$) for some families of distributions: Weibull (1), gamma (2), lognormal (3), loglogistic (4). A point (1; 2) corresponds to the exponential distribution. *a* – North Ural Bauxite Mine: I – for scale intervals with $\log_{10}E(J) < 5.0$, II – with $\log_{10}E(J) \geq 5.0$; *b* – Kamchatka-Kurils seismoactive region: I – for scale intervals with $\log_{10}E(J) < 16.0$, II – with $\log_{10}E(J) \geq 16.0$.

for these points, the coefficient of variation is near 1, as it is for an exponential distribution. The coefficient of variation for different scale levels is interesting because it characterizes the clustering of events (KAGAN and JACKSON, 1991); $\chi = 1$ corresponds to random events, and $\chi > 1$ to clustered ones. Thus even in the Catalogue of Earthquakes of North Eurasia, with removed aftershocks, weak events are clustered.

An advantage of the Weibull distribution over the exponential one is that it has an additional attenuation parameter, responsible for aftereffects or clustering; this allows the nonstationary nature of seismic processes to be taken into account. Moreover, application of the Weibull distribution has theoretical advantages. It is the limiting distribution ($n \to \infty$) for the least variable, from $n$ independent stochastic variables with the same distribution (GNEDENKO, 1962). Some models generate the Weibull distribution; for example Mukhamedov (MUKHAMEDOV, 1996) argues that the relaxation processes in a discrete hierarchy medium-like rock (the Earth's crust) obeys the Weibull distribution. The Benioff-Shimazaki model, which describes stress drop (LOMNITZ, 1994), also generates it. An additional argument for the Weibull distribution is its successful application in reliability theory.

To test the assumption of the type of the IETSL distribution, the Kolmogorov test for composite hypotheses (LEMESHKO and POSTOVALOV, 2001), with the maximum-likelihood estimation of distribution parameters, was used. The maximum allowed deviation of the empirical distribution function from the theoretical one is $\Delta F_{\max} = (6\sqrt{n}S_c - 1)/(6n)$, where $n$ is the number of elements of the analyzed sample and $S_c$ is a value which is defined by the type of tested distribution and the confidence level of the test (LEMESHKO and POSTOVALOV, 2001). It is evident that the test becomes considerably more restrictive with increasing $n$ : If the strongest considered scale level in a region corresponds to the energy class $\log_{10} E_c(J)$ with $n \approx 10$ then $\Delta F_{\max} \approx 0.32S_c$, due to the G-R law for a level with $\log_{10} E_c(J) - 2$ the number of events is about 100 and $\Delta F_{\max} \approx 0.10S_c$ and for $\log_{10} E_c(J) - 4$ $n \approx 1000$ and $\Delta F_{\max}$ is only $0.03S_c$ (for the most popular cases $S_c$ is near 1). Note that the Kolmogorov test for composite hypotheses (when the parameters of distribution are determined from the same experimental data) is more powerful than $\chi^2$-type tests.

The empirical distribution function of IETSL for different values $E$ has enough slight deviation from the Weibull distribution function not to reject the hypothesis pertaining to the Weibull distribution of IETSL with a confidence level 0.001 (the probability to reject the hypothesis about the distribution type is larger than 0.1%) generally for all energy classes, which are larger than energy class completeness threshold $\log_{10} E_{\min}(J)$ for the region considered (Table 2). With the same confidence level, the hypothesis for the gamma distribution of IETSL is rejected nearly as often as the Weibull distribution and the hypothesis for lognormal distribution is not rejected for strong events. Testing the hypothesis for the special case of the Weibull

Table 2

*Energy class over which data obey Weibull (with shape parameter k), gamma, lognormal and exponential distributions for confidence level larger or equal 0.001. $S_c$ is the parameter the of the Kolmogorov test for composite hypotheses*

| Region | Distribution | | | | | | |
|---|---|---|---|---|---|---|---|
| | Weibull | | | Gamma | Lognormal | Exponential | Completeness threshold |
| | $k$ | for non fixed $k$ | for fixed $k$ | | | | |
| $S_c$ | | 1.2784 | 1.6471 | 1.5651 | 1.6301 | 1.5593 | |
| | Energy class* | | | | | | |
| Induced seismicity | | | | | | | |
| North Ural Bauxite Mine | 0.8 | 2.2 | 2.2 | 2.2 | 4.2 | 5.0 | 3.3 |
| Gold fields Welkom (RSA) | 1.0 | 5.8 | 6.0 | 5.8 | 6.2 | 6.0 | 5.4 |
| Natural seismicity: Seismoactive regions | | | | | | | |
| Toktogul | 0.8 | 7.1 | 7.4 | 9.2 | 9.9 | 12.3 | 8.0 |
| Baikal | 0.8 | 10.3 | 10.3 | 10.3 | 11.8 | 11.2 | 10.8 |
| Kamchatka | 0.65 | 12.6 | 12.7 | 10.3 | 14.9 | 15.0 | 10.8 |
| Kamchatka-Kurils | 0.6 | 13.4 | 13.6 | 13.0 | 16.7 | 16.5 | 13.0 |

* Here and further the central value of energy class interval is written.

distribution, the exponential distribution shows that it is not rejected only for relatively strong events (Table 2). These results confirm the conclusion of the analysis shown in Figure 2.

The Weibull distribution (hazard function or failure rate function $r(\Delta t) = \lambda k \, \Delta t^{k-1}$) corresponds to the Weibull process, which is restarted after each event (BOGDANOFF and KOZIN, 1985) and which can be obtained from a simple stream by addition to the aftereffects. According to the expression for the Weibull intensity function for $k < 1$ ($k$ is shape parameter corresponding to the aftereffect), the probability of occurrence of the next event at a constant time interval decreases with time; thus events have a tendency to group (see also RYKUNOV *et al.*, 1987). Thus, the results obtained correspond to the idea concerning event interactions and therefore to temporal grouping of weak events, which leads to the occurrence of strong events (e.g., KUKSENKO *et al.*, 1996). However, the strongest events are not included in the next level of the events preparation process and have no interaction; this situation corresponds to $k = 1$ or an exponential distribution ($r(\Delta t) = \lambda_0$) with $\chi = 1$ (without aftereffect or without clustering). At the same time, it is extremely difficult to find any physical interpretation for the gamma distribution. For these reasons, the Weibull distribution was preferred.

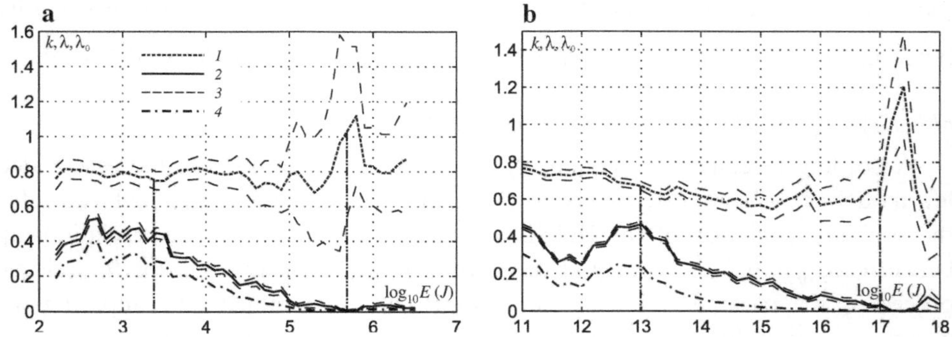


Figure 3
Behavior of the parameters of the Weibull distribution: $k$ (1) and $\lambda$ (2) with corresponding confidence intervals for confidence level 0.1 (3) and the parameter of the exponential distribution $\lambda_0$ (4) versus energy class. Interval which corresponds to the accelerated life model is marked by vertical lines: $a$ – North Ural Bauxite Mine, $b$ – Kamchatka-Kurils seismoactive region.

Figure 3 shows the behavior of the Weibull distribution parameters $k$, $\lambda$ and, for comparison, the intensity of the exponential distribution $\lambda_0$ versus scale level value. $\lambda$ and $\lambda_0$ are scale parameters which determine the intensity of events occurrence. The shape parameter $k$ corresponding to the aftereffect value is nearly a constant value and slightly less then 1. It has a gradual decreasing trend and a local increase in the area of the strongest events. At the same time, the scale parameters $\lambda$ and $\lambda_0$ decrease monotonically as the energy class grows; that is, the activity of the strong seismic events is less than that of the weak ones.

## *6. Relationships between Inter-event Times Properties and the Weibull Distribution Parameters*

The next step for IETSL investigation is to establish the restrictions on the Weibull distribution parameters to obey the accelerated life model and the G-R law. For the stochastic variable $\Delta T$, which corresponds to IETSL and has a Weibull distribution:

$$M(\log_{10}\Delta T) = -\log_{10}(\lambda e^{C})/k, \tag{8}$$

where $C \approx 0.577$ is the Euler number and

$$\sigma(\log_{10}\Delta T) = \pi \log_{10} e/(6^{0.5}\,k) \approx 0.557/k. \tag{9}$$

(Note the mean of a function of the stochastic variable $\psi(X)$ with probability density function $f(x)$ is $M(\psi(X)) = \int_{-\infty}^{\infty} \psi(x)f(x)\,dx$ and the standard deviation is $\sigma(\psi(X)) = \sqrt{M[M(\psi(X)) - \psi(X)]^2}$ (GNEDENKO, 1962).)

According to equations (8) and (9), to satisfy the conditions of the accelerated life model (5), (7), it is necessary to fix $k$, in order to have a constant value for the standard deviation of the IETSL logarithm and to have a linear relationship between $\log_{10} \lambda$ and the energy class. Therefore, the scale variable for the seismic process is the parameter $\lambda$. The same result can be obtained by comparing the general equation for cumulative distribution scaling and the expression for the Weibull distribution.

For the Weibull distribution, $T/N \cong M(\Delta T) = \Gamma(1 + 1/k)/\lambda^{1/k}$ (the value $T/N$ is a statistical estimation of $M(\Delta T)$ and $\lambda_0$), where $T$ is a considered period of seismic events registration and $N$ is the number of events on the scale level. To obtain the G-R law analogue, it is necessary to find the logarithm of the last expression:

$$\log_{10}(N/T) \cong -\log_{10}[M(\Delta T)] = (\log_{10} \lambda)/k - \log_{10}[\Gamma(1 + 1/k)]. \qquad (10)$$

Thus to obey the G-R law, it is also enough to have a constant value of the shape parameter, $k$, and a linear relationship between $\log_{10} \lambda$ and energy class.

To test the possibility of the Weibull distribution parameter $k$ being constant for all energy intervals with a maximum-likelihood estimation of $\lambda$, the Kolmogorov test for composite hypotheses has been used (LEMESHKO and POSTOVALOV, 2001). According to Table 2, the requisite coincidence between the empirical distribution and the Weibull distribution with constant $k$ parameter, generally begins at a point on the energy class completeness threshold, $\log_{10} E_{\min}(J)$. The number of strongest events for each region is not sufficient to determine the value of $k$ exactly, and for them it is also possible to have $k < 1$ corresponded to an events clustering, as for weak events. Unfortunately, variations in $k$ (Table 2) and $\sigma(\log_{10} \Delta T)$, which is defined by $k$, for different regions are too large, which means that it is not possible to write one universal scaling law for all regions.

A check on the linear relationship between $\log_{10} \lambda$ and energy class (with fixed $k$ from Table 2 and maximum-likelihood estimation for $\lambda$) was carried out simultaneously with the check on the G-R law. For this purpose, plots of values $-\log_{10}(M(\Delta T)) = (\log_{10} \lambda)/k - \log_{10}[\Gamma(1 + 1/k)]$ and $\log_{10}(N/T)$ (corresponding to the logarithm of the maximum likelihood estimation for parameter $\lambda_0$ of exponential distribution) were constructed. Both plots should be linear in representative energy intervals (intervals without loss of any events) and should coincide with each other. These properties have been observed in all regions (although the plot for the North Ural Bauxite Mine is bimodal, see above) (Fig. 4). Hence the assumption regarding the IETSL distribution self-similarity for

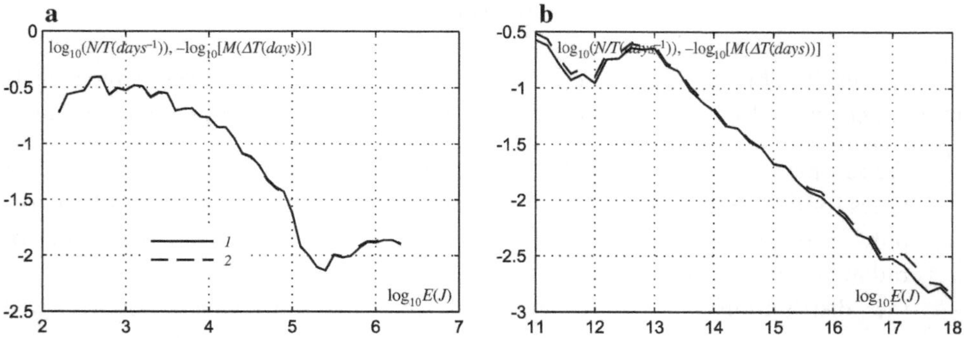


Figure 4
Comparison of the relationship $\log_{10}(N/T)$ corresponded to the G-R plot (1) with the analogue $-\log_{10}(M(\Delta T))$ written through the maximum-likelihood estimation of the scale parameter of the Weibull distribution $\lambda$ with fixed shape parameter $k$ (2), and check of the linear relationship between $\log_{10}\lambda$ and energy class. *a* – North Ural Bauxite Mine, *b* – Kamchatka-Kurils seismoactive region.

different scale levels is confirmed, with a scale coefficient $\varphi(E) = \text{const.}/E^{\gamma} = \lambda^{1/k}$. Figure 4 also shows that Eq. (10) is a good approximation of the recurrence interval plot.

To obtain a more explicit analogue of the G-R plot Eq. (8) and first best confirmed property of accelerated life model (7) should be substituted in Eq. (10):

$$\log_{10}(N/T) \;=\; -\gamma\,\log_{10} E - \delta_0 - \log_{10}\left[\Gamma(1+1/k)e^{C/k}\right], \tag{11}$$

where $\gamma$ is a parameter of the accelerated life model, which coincides with the slope coefficient of the G-R plot.

The expressions obtained for $\log_{10}(N/T)$, written with the parameters of the Weibull distribution (10) and (11), allow the reason for the common deviation from the G-R law in areas of maximum magnitude (ULOMOV *et al.*, 1999) to be inferred. Mathematically, such deviation can be related to an increase of the $k$ parameter for the strongest events. This corresponds to changes in the regime of the seismic process at this level, because these events do not prepare stronger events and at this level there is no interaction between events.

Analysis of the Weibull distribution allows the division of points in $\mu_3$ versus $\chi$ plots (see Fig. 2) into two groups with different coefficients of variation to be explained. For the Weibull distribution, $\chi(\Delta T) = \sigma(\Delta T)/M(\Delta T) = [\Gamma(1+2/k) - \Gamma^2(1+1/k)]^{0.5}/\Gamma(1+1/k)$, but for $k = 0.5\ldots1.5$, $\chi(\Delta T) \approx 1/k$. Thus a decrease in the aftereffect for strong events, with a local increase of $k$ (see Fig. 3), leads to a decrease in $\chi(\Delta T)$. This notion links well with general ideas concerning the coefficient of variation (KAGAN and JACKSON, 1991). Thus, for weak events, $\chi(\Delta T)$ is $> 1.3$, but for the strongest ones it is $\sim 1$.

## 7. Conclusions

The results obtained from the investigation of the seismic process over a scale interval covering 14 orders of magnitude of radiated seismic energy indicate that:

— the seismic process is self-similar in time over a wide range of scales: Inter-event times for different scale levels have the same type of distribution, with differences only in scale coefficient, determined by the value of this level (the value of radiated seismic energy);
— the distribution of inter-event times for different scale levels is well approximated by the Weibull distribution. At the same time, the exponential distribution approximates well the strongest events only. This fact and the observed difference in the value of the coefficient of variation $\chi$ for weak ($\chi > 1$) and strong ($\chi \approx 1$) events give evidence of the grouping tendency of weak events and the uncorrelated character of the strongest ones, but due to the statistically insignificant number of the strongest events, inter-event times of them can be described by the scaled distribution of inter-event times for weak events;
— the determined relationships allow an equation corresponding to the Gutenberg-Richter law to be derived. This confirms the consistency and validity of the results obtained;
— a detailed analysis of inter-event times, with divisions into scale levels as in the G-R law, allows information contained in the G-R plot and in the kinetics of inter-event times without division to scale levels to be combined and provides the additional possibility of strong events forecasting.

## Acknowledgements

The author is very grateful to Prof. V.A. Mansurov for the provision of induced seismicity catalogues and Hugh Rice for the language corrections.

## References

BAIESI, M. and PACZUSKI, M. (2004), *Scale-free networks of earthquakes and aftershocks*, Phys. Rev. *E 69*, 066106.

BAK, P., CHRISTENSEN, K., DANON, L., and SCANLON, T. (2002), *Unified scaling law for earthquakes*, Phys. Rev. Lett. *88*, 178501.

BOGDANOFF, J.L. and KOZIN, F., *Probabilistic Models of Cumulative Damage*, (John Wiley & Sons, New York, 1985).

CORRAL, A. (2003), *Local distributions and rate fluctuations in a unified scaling law for earthquakes*, Phys. Rev. *E 68*, 035102 (R).

CORRAL, A. (2004), *Long-Term Clustering, Scaling, and Universality in the Temporal Occurrence of Earthquakes* Phys. Rev. Lett. *92*, 10, 1–4.

CORREIG, A.M., URQUIZU, M., VILA, J., and MARTI, J. (1997), *Analysis of the temporal occurrence of seismicity at deception island (Antarctica). A nonlinear approach*, Pure Appl. Geophys. *149*, 553–574.

COX, D.R. and OAKES, D., *Analysis of Survival Data*, (Chapman and Hall, London, 1984).

DATABASE "The Catalogue of Toktogul Region Earthquakes, 1929–1991", http://zeus.wdcb.ru/wdcb/sep/toktogul/resen.html.

DAVIDSEN, J. and GOLTZ, C. (2004), *Are seismic waiting time distributions universal?*, Geophys. Res. Lett. *31*, 21, 10.1029/2004GL020892.

GARDNER, J. K. and Knopoff, L. (1974), *Sequence of earthquakes in Southern California, with aftershocks removed, Poissonian?* Bull. Seismol. Soc. Am. 64, 1363.

GERMAN, V. I., *Self-Similarity of inter-event times between failure acts in rock on different scale levels*, In Proceedings of 1-St International School-Seminar *Physical Foundation of Rock Fracture Prediction* (SibSAU, Krasnoyarsk, 2002) pp. 66–72 (in Russian).

GERMAN, V. I. (2005), *Universal scaling laws for distributions of energy, temporal and spatial characteristics in seismology*, Geophys. Res. Abstracts *7*, 01311, SRef-ID: 1607-7962/gra/EGU05-A-01311. http://www.cosis.net/abstracts/EGU05/01311/EGU05-J-01311-1.pdf.

GERMAN, V. I. and MANSUROV, V. A. (2002), *Induced seismicity monitoring and procedure of rock-burst focus localization*, J. Mining Sci. *38*, 4, July-August, 336–343.

GNEDENKO, B. V., *The Theory of Probability* (Chelsea, New York, 1962).

KAGAN, Y. Y. (1999), *Universality of the seismic moment-frequency relation*, Pure Appl. Geophysiscs *155*, 537–573.

KAGAN, Y. Y. and JACKSON, D. D. (1991), *Long-term earthquakes clustering*, Geophys. J. Int. *104*, 117–133.

KNOPOFF, L., LEVSHINA, T., KEILIS-BOROK, V. I., and MATTONI, C. (1996), *Increased long-range intermediate-magnitude earthquake activity prior to strong earthquakes in california*, J. Geophys. Res. *101*, 5779–5796.

KUKSENKO, V. S., TOMILIN, N. G., DAMASKINSKAYA, E., and LOCKNER, D. A. (1996), *A two-stage model of fracture of rocks*, Pure Appl. Geophys. *146*, 2, 253–263.

LEMESHKO, B.YU. and POSTOVALOV, S.N. (2001), *Application of the nonparametric goodness-of-fit tests in testing composite hypotheses*, Optoelectronics, Instrumentation and Data Processing *2*, 76–88. Also available at http://ami.nstu.ru/~headrd/seminar/publik_html/Awtom_eng.htm.

LOMNITZ, C., *Fundamentals of Earthquake Prediction* (Wiley, New York, 1994).

MOLCHAN, G. M. and DMITRIEVA, O. E. (1992), *Aftershock identification: Methods and new approaches*, Geophys. J. Int. *109*, 3, 501–516.

MUKHAMEDOV, V. A. (1996), *On the fractal dimension of temporal similarity in a sequence of seismic events and hierarchical structure of the crust*, Physics of the Solid Earth *31*, 6, 542–547, January, 1996. Russian Edition, June 1995. Also available at http://eos.wdcb.rssi.ru/transl/izve/9506/PAP12.HTM.

NISHENKO, S. P. and BULAND, R. (1987), *A generic recurrence interval distribution for earthquake forecasting*, Bull. Seismol. Soc. Am. *77*, 4, 1382–1399.

RAUTIAN, T. G. (1964), *About determination of the earthquakes energy for distance less then 3000 km*, Transactions of Institute of the Physics of the Solid Earth of the USSR Academy of Science *32* (199) (in Russian).

RIKITAKE, T. (1976), R*ecurrence of Great Earthquakes at Subduction Zones*, Tectonophysics, *35*, 335–362.

RYKUNOV, L. S., SMIRNOV, V. B., STAROVOIT, Yu. O., and CHUBAROVA, O. S. (1987), *Self-similarity of seismic emission in time*, Doklady AN USSR *297*, 6, 1337–1341.

SADOVSKIY, M. A., BOLKHOVITINOV, L. G., and PISARENKO, V. F., *Deformation of Geophysical Medium and Seismic Process*, (Nauka, Moscow, 1987) (in Russian).

SOBOLEV, G. A., CHELIDZE, T. L., ZAVYALOV, A. D., SLAVINA, L. V., and NIKOLADZE, V. E. (1991), *Maps of Expected Earthquakes Based on a Combination of Parameters*, Tectonophysics *193*, 255–265.

SORNETTE, D. and KNOPOFF, L. (1997), *The paradox of the expected time until the next earthquake*, Bull. Seismol. Soc. Am. *87*, 4, 789–798.

Special Catalogue of Earthquakes of the Northern Eurasia, ancient time-1990 (ed. Kondorskaya, N.V. and Ulomov, V.I.), http://www.seismo.ethz.ch/gshap/neurasia/nordasiacat.txt.

TOMILIN, N. G. and VOINOV, K. A., *Technique and results of the rock burst prediction* Proc International Conf. Mechanics of Jointed and Faulted Rock, (Rotterdam: Balkema, 1995) pp. 955–959.

TURCOTTE, D. L., *Fractals and Chaos in Geology and Geophysics* (Cambridge University Press, Cambridge, U.K., 1992).

ULOMOV, V. I. (1998), *Focal Zones of Earthquakes Modeled in Terms of the Lattice Regularization* Izvestiya, Physics of the Solid Earth *9*, 717–733.

ULOMOV, V., SHUMILINA, L., and TRIFONOV, V. *et al.* (1999), *Seismic hazard of northern eurasia,* Annali Geofis. *42*, 1023–1038.

WANG, J. H. and KUO, C. H. (1998), *On the frequency distribution of interoccurence times of earthquakes,* J. Seismol. *2*, 351–358.

(Received March 11, 2005, revised October 30, 2005, accepted November 17, 2005)
Published Online First: September 14, 2006

GPSR Compliance

*The European Union's (EU) General Product Safety Regulation (GPSR) is a set of rules that requires consumer products to be safe and our obligations to ensure this.*

*If you have any concerns about our products, you can contact us on ProductSafety@springernature.com*

In case Publisher is established outside the EU, the EU authorized representative is:

Springer Nature Customer Service Center GmbH
Europaplatz 3
69115 Heidelberg, Germany

**Batch number: 10584556**

Printed by Printforce, the Netherlands